P9-CMY-436

ERUPTION

ERUPTION

THE UNTOLD STORY OF MOUNT ST. HELENS

Steve Olson

W. W. NORTON & COMPANY
Independent Publishers Since 1923
NEW YORK • LONDON

Printed in the United States of America
First Edition

For information about special discounts for bulk purchases, please contact
W. W. Norton Special Sales at specialsales@wwnorton.com
or 800-233-4830

Manufacturing by RR Donnelley, Harrisonburg, VA
Book design by Daniel Lagin
Production manager: Anna Oler

Library of Congress Cataloging-in-Publication Data

Names: Olson, Steve, 1956–
Title: Eruption : the untold story of Mount St. Helens / Steve Olson.
Description: First edition. | New York : W.W. Norton & Company, [2016] |
Includes bibliographical references and index.
Identifiers: LCCN 2015038842 | ISBN 9780393242799 (hardcover)
Subjects: LCSH: Saint Helens, Mount (Wash.)—Eruption, 1980. | Volcanic
eruptions—Washington (State) | Volcanoes—Washington (State)
Classification: LCC QE523.S23 O47 2016 | DDC 363.34/95—dc23 LC record
available at http://lccn.loc.gov/2015038842

W. W. Norton & Company, Inc.
500 Fifth Avenue, New York, N.Y. 10110
www.wwnorton.com

W. W. Norton & Company Ltd.
Castle House, 75/76 Wells Street, London W1T 3QT

1 2 3 4 5 6 7 8 9 0

For Rafe, for his guidance, encouragement, and friendship

And for Sid and Adele, for their steadfast support

CONTENTS

PROLOGUE

> Natural disasters are revelatory. The manner in which a society interprets a catastrophe and responds to the chaos exposes many of the accepted truths, prejudices, hopes, and fears of a culture.
>
> —Nicholas Shrady, *The Last Day*

IN THE YEAR 1980, TOWARD THE END OF MARCH, NEWSPAPERS AND television stations began reporting on a series of strange events taking place in a largely unknown corner of the western United States. A volcano in southwestern Washington State known as Mount St. Helens was threatening to erupt. A crater had opened on the summit of the mountain and was spewing ash and steam thousands of feet into the air. Earthquakes were shaking the volcano so violently that people nearby said it was like being on a ship at sea. State officials, based on geologists' predictions, were telling residents and visitors to stay away. Floods, mudflows, and withering blasts of superhot gas could, with little warning, sweep away trees, houses, and people.

It was a wonderful diversion at an unhappy time in the nation's history. In 1980 the United States was still recovering from the traumas of

Vietnam, Watergate, and the oil embargos of the 1970s, which had temporarily deprived Americans of one of their most cherished freedoms—the right to drive anywhere, anytime, for as long as a person might want. A long presidential campaign was just getting under way between an unpopular sitting president, Jimmy Carter, and the eventual Republican nominee, a former California governor and movie star named Ronald Reagan, who promised to return the nation to its former glory. Students backed by the revolutionary government of Iran were holding fifty-two Americans hostage in the US embassy. The most popular music of the time was disco; fashions ran from bell-bottoms to peasant blouses; and men sported bushy mustaches and long sideburns. In a 1976 magazine article, Tom Wolfe referred to the 1970s as "The Me Decade" for the period's pervasive dissatisfaction and devotion to personal transformation, and the label seems more appropriate than any decadal label coined since.

In the Pacific Northwest, the mood was even grimmer. In the 1970s, when I left Washington State to go east for college, the region seemed to me to be teetering on the edge of insolvency. Employment at Boeing, the state's largest company, had dropped from more than 100,000 people to less than 40,000, prompting two real estate agents to erect the famous billboard near Seattle-Tacoma Airport that read WILL THE LAST PERSON LEAVING SEATTLE—TURN OUT THE LIGHTS. A scrappy band of troublemakers, emboldened by the environmental movement of the 1970s, was threatening the state's second largest industry—the extraction of timber from the vast forests of the Cascade Mountains. Commercial and sports fishermen were complaining about declining catches—the product of too much development, too many dams, and too much fishing. The copious bounty of the Northwest—celebrated in the potlatches of the native peoples, in nineteenth-century railroad advertisements, in the breathless promises of civic boosters, and in countless middle-class households like my own—seemed in danger of fading away.

In the midst of all this economic, geopolitical, and cultural gloom, a misbehaving volcano took people's minds off their troubles. It made at least one small part of the country seem exotic, mysterious, exciting. After all, volcanic eruptions did not occur in the United States—at least not outside the distant annexes of Alaska and Hawaii. Though hurricanes and tornadoes still struck the South and Midwest with alarming frequency, the last volcano to erupt in the contiguous United States had been Lassen Peak in Northern California in 1917, so long ago and in such an isolated part of the country that it was all but forgotten. The last earthquake in the contiguous United States to cause more than 100 fatalities, in Long Beach, California, had occurred almost a half century before, in 1933. The United States felt settled and mundane, as if Americans had tamed not just the landscape but the very geological foundation on which their homes, roads, and businesses stood. That a formerly tranquil mountain in an otherwise unassuming range of peaks could suddenly spout fire seemed almost miraculous.

Less than two months after it first rumbled to life—on a bright Sunday morning in May—Mount St. Helens exploded with a violence out of all proportion to what it had done before. It was the single most powerful natural disaster in US history—more powerful than Hurricane Katrina when it hit New Orleans in 2005, more powerful than the 1906 earthquake that destroyed San Francisco, more powerful even than the world's first atomic explosion on the high plains of New Mexico in 1945. The eruption generated one of the largest landslides in recorded human history; it devastated an area four times the size of Washington, DC; and it killed people thirteen miles away from the volcano's summit—the distance from Central Park to Coney Island, or from downtown Los Angeles to the Pacific Ocean. Photographs taken from satellites a few weeks after the eruption show an angry gray blot amid the jade-green forests of southwestern Washington State, as if a gigantic hand had swept the earth bare.

A natural disaster is not a disaster until it becomes a human disaster; otherwise, in the minds of most people, it is mere spectacle. Fifty-seven people died that Sunday morning—a number that would have been ten times higher if the volcano had exploded on a weekday rather than on a Sunday. The dead were swept off hillsides, crushed by falling trees, carried away by floods, asphyxiated by ash. The bodies of almost half were never found and remain buried around the mountain.

I became interested in writing this book while thinking about those fifty-seven people. What led them to be where they were that fatal Sunday morning? After all, many people in the Northwest could have been numbered among the dead. A friend could have called and said, "Let's go see the volcano," and a group of two or three or four would have climbed into a station wagon or RV, driven down Interstate 5, turned off on the Spirit Lake Highway, and camped overnight on a ridge with a good view of the mountain. When people who were in the Northwest in 1980 remember where they were the day the volcano erupted, many do so with the prickly sensation of having barely escaped disaster.

I also wondered about the legacy of those who died. In the Northwest and elsewhere, they often are remembered as daredevils who violated the law to get where they were when the volcano exploded. But that belief is completely at odds with the truth. Why do so many unfair misconceptions surround their deaths?

In retrospect, the fifty-seven people who died were too close to a dangerous volcano. That always struck me as odd. People have known for millennia that volcanoes can be deadly. On August 24 in the year 79 CE, after centuries of inactivity, the volcano known as Vesuvius, halfway down the west coast of Italy, suddenly erupted an immense column of ash and smoke that towered high above what is now the city of Naples. The scientist, author, and naval commander Gaius Plinius Secundus, known to history as Pliny the Elder, was in his villa in the town of Misenum, twenty miles to the west of Vesuvius across the Bay of Naples, when the volcano erupted. He immediately ordered that a boat be prepared so he could investigate the eruption. He died about ten miles south of the volcano, either from inhaling toxic gases or from a heart attack brought on by his exertions. In the coastal town of Herculaneum, four miles from the summit, and in Pompeii, six miles from the summit, thousands more died when their towns were overrun by volcanic blasts and buried by hurricanes of hot ash.

The day of Pliny the Elder's death, his sister and her son, Pliny the Younger, tried to flee Misenum as a cloud of volcanic ash expanded toward the town. Later, Pliny the Younger wrote about their experiences on a nearby hill:

> It was not the darkness of a moonless or cloudy night, but just as if the lamp had been put out in a completely closed room. You could have heard women shrieking, children crying, and men shouting. Some were calling for their parents, some for their children, some for their wives, and trying to recognize them by their voices. Those people were bewailing their own fate, or those of their relatives. Some people were so frightened of dying that they actually prayed for death. Many begged for the help of the gods, but even more imagined that there were no gods left and that the last eternal night had fallen on the world.

Today, eruptions like that of Mount St. Helens, with columns of ash that tower far overhead and descend upon their victims like the angel of death, are still known as Plinian eruptions to honor Pliny the Younger's careful descriptions of the event.

• • •

To understand why fifty-seven people were killed in the eruption of Mount St. Helens, I first had to learn more about the geology of volcanoes. Geologists had obviously underestimated how large the eruption could be. But they had plenty of warning: Mount St. Helens had been shaking, smoking, and swelling for nearly two months before the cataclysmic eruption of May 18. How could they have been so surprised when it exploded?

I soon learned something else. Geological factors were only part of the reason for the disaster. The people who were around Mount St. Helens had many reasons for being where they were. Some of those reasons extended just a day or two into the past—to a promising weather forecast, to the opening of fishing season, to an incautious curiosity. But other reasons went back decades or centuries—to a lumberyard on the banks of the Mississippi River, to a dining room on Summit Avenue in St. Paul, to a railroad snaking across the high plains. Great forces that have shaped the nation throughout its history—the extraction of natural resources, the construction of nationwide transportation systems, the rise of the conservation movement—helped determine the fates of the people scattered around the volcano that Sunday morning.

To my surprise, a major theme of the story was logging—though perhaps I shouldn't have been surprised, given the importance of logging in the Pacific Northwest. Throughout the nineteenth and twentieth centuries, thousands of men and their families streamed into the states of Washington, Oregon, and Idaho to cut down the immense trees that once blanketed the hills and valley floors of the region. My own great-grandfather, who was born and raised in Boston, was killed in the forests east of Seattle when a massive tree he was loading onto a truck fell off and crushed him. My grandfather, who was helping him on the job, had to wrap his body in a blanket and drive it back over the mountains to my great-grandmother. My father worked summers during high school and college at a gigantic sawmill in Everett, a half hour north of Seattle, before opting for the more prosaic life of an accountant. Sawdust runs in the veins of many Pacific Northwest families.

Many companies have logged the great forests of the West, but one in particular looms large in the story of Mount St. Helens. In the year 1900, the lumberman Frederick Weyerhaeuser, who already had made a fortune cutting down the trees of Wisconsin and Minnesota, bought substantial portions of the land between Mount St. Helens and the Pacific Ocean, sight-unseen, from his next-door neighbor on Summit Avenue in St. Paul, Minnesota. That purchase turned out to include some of the richest and most productive forestland anywhere in the world. What John Muir wrote of America's forests applied especially to those of southwestern Washington: "The forests of America, however slighted by man, must have been a great delight to God; for they were the best he ever planted."

The forests west of Mount St. Helens made the Weyerhaeusers, already a rich family, an American dynasty. Early in the twentieth century, many members of the family and their business associates relocated from the largely cutover Midwest to the virgin forests of the Pacific Northwest. There they sought to blend in with the reserved and self-effacing culture of their new homeland. But even today, in the upper-left-hand corner of the country, no family name is as richly evocative as Weyerhaeuser.

It is difficult to overstate the significance of the Weyerhaeuser Company to the history of the Pacific Northwest. William Boeing dropped out of Yale and moved to Washington State in 1903 to become a lumberman when he learned about Frederick Weyerhaeuser's massive timber purchase—only later realizing that the gigantic spruce trees that grew in Washington's soggy loam yielded ideal frames for the wonderful new invention of airplanes. Whole towns and cities took shape around Weyerhaeuser's operations, thriving if that part of the business succeeded or withering if the business faltered. The company pioneered the idea of tree farms that would grow successive crops of trees in perpetuity. The physical appearance of the land, on drives through the Pacific Northwest hinterland, was often the result of decisions made in Weyerhaeuser's headquarters.

Weyerhaeuser and other economic interests have formed the backdrop against which much of the region's history has played out. That turned out to be the case at Mount St. Helens too.

• • •

Another reason I wanted to write this book is because 1980 was such a pivotal year in American history. Before 1980, income disparities in the United States were narrowing; since then, incomes have become steadily more unequal. Union membership peaked in 1979 and has never been as high since. The year 1980 saw the highest divorce rate in US history. The release of carbon dioxide into the atmosphere accelerated after 1980 as more countries joined the United States and Europe in burning fossil fuels without restraint.

The year 1980 was even more consequential in the Pacific Northwest. In 1979, Bill Gates and Paul Allen moved their fledgling company, then named Micro-Soft, from Albuquerque, New Mexico, back to their hometown. In 1982, Howard Schultz became the marketing director for the old downtown Starbucks coffee shop, which he bought a few years later and began to expand. In 1983, Costco opened its first store in south Seattle and quickly became the fastest-growing company in US history.

A few years after that, Jeff Bezos founded Amazon after driving to Seattle while drawing up cost projections for his new venture on the way. Today, those four companies—Microsoft, Costco, Starbucks, and Amazon—are the largest and most profitable companies in the state. Logging, fishing, and agriculture, while still active, have faded into the background.

Of course, the eruption of Mount St. Helens did not bring about the economic revival of the Pacific Northwest, at least not directly. If anything, people moving to the Northwest worried for a few years about relocating near an active volcano. But the eruption focused attention on a part of the country that previously had been an afterthought. The Pacific Northwest has a combination of assets unknown elsewhere—a breathtaking natural beauty, a gentle rain-washed climate, a tolerant and ecumenical culture. As the economies of Seattle, Portland, and other big cities boomed in the 1980s, the media began to celebrate the region's innocence and quirkiness—from the Space Needle to Pike Place Market, from *Sleepless in Seattle* to *Frasier*. By 1996, in a cover story entitled "Seattle Reigns," *Newsweek* proclaimed, "Sooner or later, it seems, everyone moves to Seattle, or thinks about it."

At the same time, the revival of cities in the Pacific Northwest masked a darker trend—one more accurately captured by *Twin Peaks* than *Frasier*. In small towns like the one where I grew up, the decline of the old-time extractive industries and the cultural fallout of the 1960s and '70s caused immense human hardship. The incomes of men began to fall, and more women entered the workplace. Rising divorce rates caused families to dissolve and reform, leaving children confused and directionless. Illicit drugs, from cocaine in the 1980s to meth in the 2000s, hollowed out the lives of friends and spouses and cast a pall over small-town life. Of course, these trends were not limited to small towns, and many people remained in small towns or moved to them to escape the stresses of city life. But small towns suffered disproportionately.

The eruption of Mount St. Helens marked the dividing line between the old Northwest and the new, between the decline of the countryside and the rise of the cities, between an economy based on resources and one based on ideas. The stories of the people who were around the mountain when it exploded reflect this turning point, as if caught in an unexpected snapshot.

• • •

Finally, the 1970s and '80s were decades of heightened environmental activism, which also is part of the Mount St. Helens story. The first Earth Day was held on April 22, 1970, and the environmental movement grew in size and influence for the next decade. The *Limits to Growth* report released by the Club of Rome in 1972 suggested that civilization would collapse in the twenty-first century if consumption continued unchecked. In about 1980 the first articles began to appear in national magazines warning that the continued burning of fossil fuels could cause the climate to warm. A few years later the nation went through a brief frenzy at the thought that even a limited nuclear war could plunge the planet into a prolonged nuclear winter.

In the midst of this upwelling of environmentalism, the eruption of Mount St. Helens added a note of uncertainty, of peril, to how people thought about the future. It reinforced the idea that the earth is at best indifferent and at worst hostile toward its human occupants. The eruption of the volcano took no notice of human affairs. It snuffed out the lives of people who were too close to the mountain as casually as a leaf drops from a branch. The people who experienced the eruption could never again think of the ground on which they stood as beneficent and forgiving. The earth could reach up and crush them at any time.

The loss of life at Mount St. Helens occurred because people were unable to see a risk that in retrospect was obvious. They thought that the risk was small, or that they were smart enough to get out of the way if something happened. Most of us go through life this way. We ignore the risks we face so we are not paralyzed by dread. Only in retrospect does the extent of our willful ignorance become clear.

PART 1
THE LAND

Mount St. Helens before the 1980 eruption, with logged Weyerhaeuser land and Fawn Lake in the foreground

THE AWAKENING

AT 3:47 IN THE AFTERNOON ON MARCH 20, 1980, IN THE BASEMENT of the Geophysics Building on the campus of the University of Washington, not far from where Interstate 5 slices through the hills and waterways of downtown Seattle, the stylus of a seismic recorder leapt across a slowly rolling drum of paper. Somewhere south of Seattle, beneath the wooded hills and glacier-clad peaks of the Cascade Mountains, an earthquake was under way.

A few minutes later the seismology lab's data analyst, Linda Noson, walked by the recorder and glanced at the display. Wow. That had to be at least a four. She hadn't seen an earthquake that big in the state for almost a year.

In his office upstairs, Steve Malone was talking with Craig Weaver when Noson walked in. "We just got a four from the Mount Rainier station," she told them.

Back downstairs, the three of them peered at the jagged line of the seismogram.

"I bet it's Mount Hood," said Weaver. "I've been waiting for that."

"I'm thinking it's St. Helens," said Malone.

Why speculate when you can figure out the answer?, Noson thought. As the men left the room, she took a ruler from a desk and measured on the printouts the exact moment when the shaking arrived at each of several seismic stations in the state. Then she entered the results onto computer punch cards and took the stack upstairs to send the data to the campus computer center—the same computer center where, a few years earlier, a sixteen-year-old Bill Gates and his friend Paul Allen had hung out to cadge time on the mainframe. She sat down to read a magazine while the computer crunched the numbers.

A few minutes later, she walked back into Malone's office.

"Steve's right," she said. "It's Mount St. Helens."

After 123 years of peaceful slumber—years in which railroads and freeways had replaced the Indian trails of previous centuries, in which towns and then cities had sprung up next to rivers and saltwater inlets, in which the primordial forests of the Pacific Northwest were sawed down and made into railroad ties, boats, planes, paper, homes, and office buildings—Mount St. Helens had woken up.

• • •

A week later, Dave Johnston stood in a parking lot on the north slope of Mount St. Helens and grimaced at a phalanx of camera-toting reporters. He hated doing interviews, but if this was what it took to be near the volcano, he would put up with it. He looked more like a lumberjack than a geologist. He wore a blue stocking cap, red-checked flannel jacket, black turtleneck, and scuffed boots. He was clean-shaven, thirty years old, blond, handsome in a gangly kind of way. As the shutters of cameras clicked in the bright sunshine, he turned and squinted at the mountain. "This is an extremely dangerous place to be," he told the reporters. "If it were to explode right now, we would die."

Dave Johnston was an unusual choice for the first scientist to make a prominent public statement on Mount St. Helens's reawakening. He had received his PhD just two years before and was an expert on the gases given off by volcanoes, not their seismic activity or eruptive potential. But Johnston loved volcanoes, and he had a knack, with his enthusiasm and friendliness, for getting himself in the middle of the action. He had grown up in Oak Lawn, Illinois, a largely blue-collar suburb southwest of Chicago. The son of an engineer father and newspaper editor mother, he wanted nothing more in high school than to be a *National Geographic* photographer. But a C in English at the University of Illinois dimmed his enthusiasm for journalism. Instead he began taking geology courses and was immediately hooked. He did fieldwork one summer mapping the ancient volcanoes of southwestern Colorado, and from then on volcanoes were his passion. In 1971 he moved to Seattle to enter graduate school for geology at the University of Washington, where he eventually began studying volcanic gases. Molten rock near the surface of the Earth gives off a variety of compounds, including carbon dioxide, sulfur dioxide, and hydrogen sulfide. In 1980 Johnston was working for the US Geological Society to determine whether volcanic eruptions could be predicted by monitoring the gases they emitted, which is one reason why he was helping to keep tabs on Mount St. Helens.

Volcanologists have a tendency to drift westward in the United States because that's where the action is tectonically. North and Central America occupy the western portion of a big slab of the earth's crust known as the North American plate, which is shaped roughly like an inverted triangle. The bottom of the triangle is in the middle of the Atlantic Ocean halfway between South America and Africa. The top two corners are north of Siberia and northwest of Greenland. This piece of the earth's crust is constantly jockeying for position with the tectonic plates that surround it. In some places, like Iceland, the North American plate is pulling away from an adjoining plate, and molten material is welling up to fill the gap. In other places, like California, the North American plate is slipping past an adjoining plate, often getting stuck and then breaking free in earthquake-inducing jolts.

But the most dramatic and dangerous of these plate interactions occur in the Pacific Northwest. There, in a line from southern British Columbia to Northern California, a small piece of oceanic crust is being forced under the edge of the North American plate at the rate of a few inches per year. As this crust descends into the earth, it and the oceanic water it carries get hotter, which causes the surrounding rock to melt. This molten rock is more buoyant than the surrounding solid rock, so it rises through cracks in the earth, melting additional material on the way, until it reaches the surface. For millions of years, this recycled oceanic crust has been erupting and building up immense peaks that rise thousands of feet above the surrounding countryside. These western volcanoes are so high that the summer sun cannot completely melt the snow on their flanks. Gradually the snow compresses, forming glaciers that over the millennia have gouged out jagged valleys in the smooth volcanic cones. About every fifty miles in southwestern Canada and the northwestern United States, from Mount Garibaldi in Canada to Lassen Peak in California, great glaciated peaks rise above the surrounding mountains, hooded sentinels above a green-clad landscape.

Because of the tectonic activity occurring beneath the West Coast, earthquakes are not unusual in the Pacific Northwest. Every decade or two the region gets a good shaking, and huge earthquakes occur every few centuries. But usually the earth quiets down after an earthquake until the pressure to move again becomes unbearable.

After the first earthquake beneath Mount St. Helens, the earth did not quiet down. On the contrary, it got more agitated. By the following Tuesday, five days after the initial earthquake, hundreds of earthquakes had occurred beneath Mount St. Helens, often following one another so closely that they ran together on the seismometers. That day, the Forest Service closed the mountain above the tree line to climbers and skiers so they wouldn't be buried by snow avalanches. But overflights during rare moments of good weather showed the pristine white cone of the mountain still to be unsullied.

Right after lunchtime on Thursday, people near Mount St. Helens heard a loud bang from the cloud-covered volcano. A few hours later a reporter from a Portland radio station was flying over the mountain when the clouds suddenly parted and revealed a plume of steam and ash rising from the mountain's summit. "There is no question at all," he radioed to his listeners. "Volcanic activity has begun. You can see smoke and ash pouring from the top of the mountain, especially the north side of the mountain." A blackened crater 250 feet across had opened on the top of the mountain and was showering ash on the mountain's northeast side. Officials from the Washington Department of Emergency Services told everyone within fifteen miles of the volcano that they should leave the area.

The area around Mount St. Helens is not heavily populated—at least not at the end of March, when the lakes surrounding the mountain are still frozen and the ridgelines remain covered by snow. All around the volcano, narrow valleys, steep hillsides, and dense forests have discouraged human settlement. To the northwest, the nearest town is Toutle, twenty-eight miles away. (The name is from a band of Native Americans who lived in the area before the arrival of American settlers.) To the southwest, the nearest towns are Cougar and Yale, which serve mostly to provision people working, camping, fishing, and hunting in the nearby forests. East of the volcano the country is even more wild and is mostly included within the Gifford Pinchot National Forest.

But the area around Mount St. Helens was not devoid of people when the volcano began shaking. Every weekday, hundreds of loggers employed by the Weyerhaeuser Company were working in the woods around the mountain. Weyerhaeuser owned most of the land between Mount St. Helens and Interstate 5, thirty-five miles to the west, and for the previous eight decades it had been logging that land hard. Most of the original forests were gone, replaced by muddy clearcuts and even stands of hand-planted trees. But right around Mount St. Helens some of the old-growth forests still stood—gigantic trees 10 or more feet across and more than 250 feet tall, monsters that shook the ground so hard when they fell that loggers nearby could barely remain standing. In the spring of 1980, Weyerhaeuser was chopping down its last remaining stands of these trees. To the west and northwest of the volcano, the woods were filled with logging roads and yarding towers.

When the volcano erupted Thursday afternoon, Weyerhaeuser evacuated three hundred of its employees from the area. For many loggers, it was a chance to retreat to their favorite bars in Toledo, Vader, and Castle Rock an hour or two early. At the Hill Billy Inn, the girlfriend of a logger was quoted in the *Portland Oregonian* saying, "The mountain is blowing, and this tavern is going." But their time off was brief—as one company official scoffed, "There's no concern of any immediate danger at all." By the next morning, they were all back at work.

Thursday morning, a television station in Seattle had offered Steve Malone a helicopter ride to the volcano in exchange for an interview. Malone was too busy, so he asked Johnston if he wanted to go. Johnston was so nervous about public speaking that he once hyperventilated and passed out while giving a scientific talk, but this offer was too good to turn down. He drove to the helicopter pad, shook hands with the journalists, and took off.

The helicopter carrying Johnston and the reporters landed in the Timberline parking lot on the north flank of the volcano. Timberline was the end of the road to Mount St. Helens—a broad paved area just above the volcano's highest grove of trees for drivers who wanted to get as close to the summit as possible. From the parking lot, nothing but glaciers and snowfields stood between the reporters clustered around Johnston and the mountain's peak. The afternoon was bright but cold. Wind-driven snow needled their faces.

Johnston was well aware of the risks they were taking. He had done much of his graduate work on Mount Augustine, an exploding volcano on an uninhabited island southwest of Anchorage. One time he and six other volcanologists had flown to the island to study the volcano, and a fast-rising storm had wrecked their helicopter shortly after they landed. For three days, Johnston and his colleagues huddled in a corrugated-metal shack, with canned peaches for food and a little jet fuel for heat. Twelve hours after they were rescued, the volcano exploded again. Temperatures in the shack got hot enough to melt plastic. No one would have survived.

"Magma is rising," he told the assembled reporters in the parking lot beneath Mount St. Helens. "It looks like there's a very good chance there will be an eruption. If there is an explosion, it is possible that very, very hot incandescent debris could come down on all sides." To the reporters filming Johnston and taking down notes, the incongruity of what he was saying was obvious. He was talking about all of them dying in a devastating eruption. Yet he was obviously excited to be so close to an active volcano. The reporters were excited too—it was a great story, and there was something exhilarating about witnessing a geological process so large, dramatic, and rare. Johnston shielded his eyes with one hand, turned to gaze at the volcano, and grinned. "We are standing next to a dynamite keg, and the fuse is lit," he said. "We just don't know how long the fuse is."

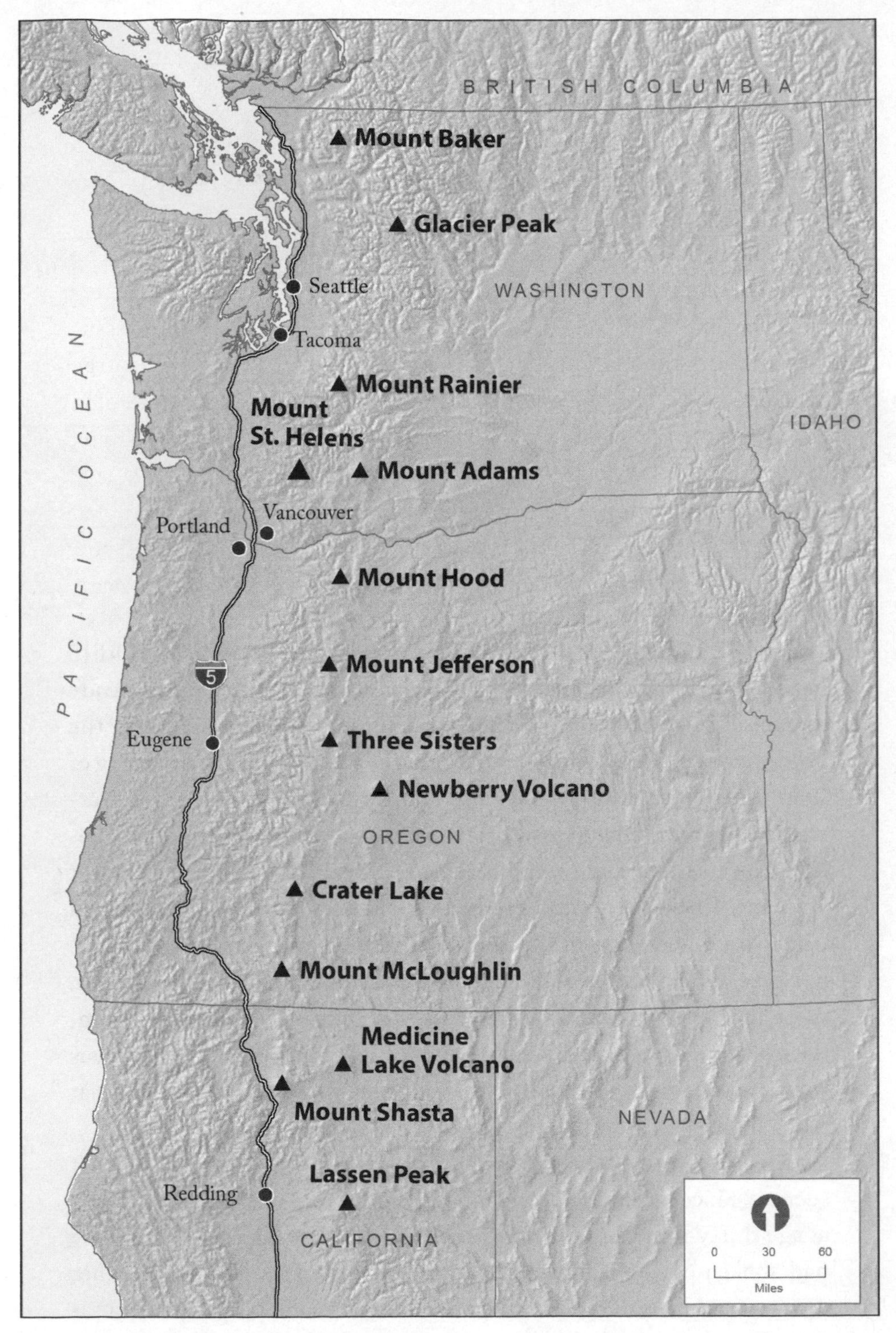

The volcanoes of the western United States

THE VIEW FROM THE EXECUTIVE SUITE

FROM THE TOP FLOOR OF THE WEYERHAEUSER COMPANY'S ICONIC headquarters building, halfway between Seattle and Tacoma, a steady stream of cars can be seen heading north on Interstate 5, just on the other side of a kidney-shaped pond. This is where, in the early spring of 1980, the Weyerhaeuser Company and its president, George Weyerhaeuser, decided how to respond to the reawakening of Mount St. Helens. The company was well positioned to deal with the disruption. It was one of the largest wood-products businesses in the world. It owned nearly 6 million acres of the United States—an area larger than Connecticut and Rhode Island combined—and had rights to log on even more land than that in Canada, Malaysia, Indonesia, and other countries. On the basis of its extensive timber holdings, Weyerhaeuser was the richest company in Washington State—richer even than Boeing. It had a higher market valuation than Ford, Mobil, or Xerox.

George Weyerhaeuser was the man responsible for much of that success. Since becoming president and chief executive officer in 1966—at age thirty-nine, the youngest president in Weyerhaeuser history—he had shaken up the company in ways that his father and grandfather never would have dared. He had accelerated the cut on company lands and plowed the profits back into land, sawmills, and paper plants. He had cleared out the deadwood from the executive suite; people still spoke in hushed tones of that bloodletting. He had diversified into new businesses. By 1980, Weyerhaeuser was the United States' fifth largest homebuilder. Its mortgage subsidiary was the fourth largest mortgage banker in the nation. The company ran a salmon-ranching business in Oregon, freshwater shrimp farms in Florida and Brazil, and a wholesale nursery supply business. Weyerhaeuser was a Forbes 100 company in the United States and a force to be reckoned with anywhere in the world.

At first glance, one would not expect Weyerhaeuser to be such a forceful leader. A compact man, he had brown eyes, a prominent forehead, and curly brown hair that seemed incongruous in a chief executive officer. He wore dark, heavy glasses, chosen obviously to be functional rather than stylish. His suit and tie were off the rack, his shoes unpolished. He drove himself to work or took the company van with his employees. For vacation, he put his six kids in the car and drove them to Disneyland, where they stayed in a modest hotel.

But a closer look yields a different impression. Weyerhaeuser had been a boxer in high school and during his navy training, and as an adult he always carried himself like a boxer—chin forward, chest up, eyes searching for an opening. As one interviewer described him, "There is a quiet power behind his gestures. He speaks like a man who is accustomed to having his way." When talking with someone, he turned to face that person directly, so that his manner was uncommonly open, even confrontational. He did not shy away from anyone or anything, even when something was troubling him.

And Weyerhaeuser had plenty to worry about in the spring of 1980. This business with the volcano couldn't have happened at a worse time. The housing market was collapsing. A few months earlier President Carter had appointed Paul Volcker to head the Federal Reserve, and Volcker had vowed to do whatever it took to squeeze inflation out of the economy. With the prime at 20 percent, housing starts had dropped by almost half from the year before, with no upturn in sight. The timber industry was headed toward another contraction, just like all the ones that had come before, just like the ones that had been threatening the company for more than a hundred years. There seemed to be no way to rid the industry of these endless crashes. Being diversified helped, but it would only reduce the pain.

Weyerhaeuser had no good options. He could lay off more employees, but thousands were already out of work. He could curb some of his other businesses, but that might make some of the recent acquisitions look bad. He had to cut fewer trees, but that would eat into revenues. The lumber business was built on a contradiction—the more the company needed to cut and sell those trees, the less people wanted to buy them.

But there was one thing he couldn't do. He couldn't quit logging around Mount St. Helens. It was the best land the company had ever owned. His great-grandfather had no idea what he was getting when he bought that land. Some of the old-timers said that it was the most productive timberland anywhere in the world. For almost eighty years the company had been logging between Interstate 5 and Mount St. Helens. It had extracted billions of dollars from those forests. Entire towns had grown up around the logging business. Families had worked for the company for generations.

The big trees were almost all gone now, the ones large enough to build an entire neighborhood with a single fall. Weyerhaeuser would finish logging the old growth and close the old mills. The old-growth timber the company had owned would be a relic of an irretrievable past.

And then, as if the volcano weren't enough, there were the environmentalists to deal with. They wanted to build a park around Mount St. Helens that would include the company's land, the land his great-grandfather had bought fair and square from Jim Hill eighty years earlier. If Weyerhaeuser let that happen, where would it stop? His company was the largest landowner in the state. If the state or the feds got the idea that they could tell Weyerhaeuser what the company could do on its own land, they could lock up those trees forever. That would be the end of it. Four generations of Weyerhaeusers and the company would end on his watch.

It wasn't as if George Weyerhaeuser didn't know what those forests were like. He had worked as a choker setter in the woods during college.

He had walked among those trees, their limbs climbing toward heaven, where the light filtered through the canopy and caught in the undergrowth, as if time itself were holding its breath. Anyone who has worked as a logger can remember that moment, the fading buzz of the chainsaw, and the sudden slackening as the tree begins to fall, the astonishing sense that something so large could succumb to the efforts of a single handheld tool, followed by that sound, the crackling of a falling tree, like a bolt of lightning at close range, as the moment grew to the inevitable explosion. Bringing daylight to the swamp, the old-timers called it. But the swamp didn't go on forever. Even the old-timers were starting to realize that.

Of course, the environmentalists would complain when the old growth was gone. But the outcome was preordained. Already most of Weyerhaeuser's land consisted of tree farms. Timber was a crop—the company had been saying that for fifty years. Trees grew like weeds in the Pacific Northwest. Cut them down and they grew right back—maybe not in a single lifetime, but in the lifetime of a man's sons, or of his future grandsons. That's what Weyerhaeuser's great-grandfather had said not long after founding the company: "This is not for us, nor for our children, but for our grandchildren."

Besides, if Weyerhaeuser wasn't an environmentalist, who was? He had kept the company's land open to hunting and fishing and camping—despite the damage people did, the trash and stealing, the ever-present threat of fire. He was probably responsible for the planting of more trees than any other person in human history. If the environmentalists thought Weyerhaeuser was bad, they should see what the other companies did to the woods. Yet for all the company's efforts to be a good citizen, it was known among environmentalists only as "the best of the S.O.B.'s." Which meant that Weyerhaeuser was what exactly?

Plenty of old growth was left—in the national parks, some in the national forests, even on a few pieces of private land. But the trees would deteriorate, like everything else. They would fall someday, rot, and return to the earth from which they came. It was a waste to let those trees die. In a world without humans, it might have made sense. In the modern world, it made no sense at all.

And now this volcano—it made everything more complicated. He had to get those trees out. He could talk with the governor. He was a member of her kitchen cabinet and had given plenty to her campaign. She would be sympathetic to his problem.

When Weyerhaeuser's headquarters opened in the early 1970s, it had won architectural awards for the way it blended into the landscape. It stretched between two hillsides—a groundscraper, people called it—with ivy growing between each of its terraced floors. It almost looked like part of the land, as if it had been carved out of the hillside.

On the top floor of his company's headquarters, eighty miles north of Mount St. Helens, George Weyerhaeuser couldn't feel the earthquakes that passed, like a trembling leaf, beneath his feet.

THE FLOOD

ONE HUNDRED YEARS EARLIER, FREDERICK WEYERHAEUSER emerged from his home in Rock Island, Illinois, and beheld, with great satisfaction, a town built almost entirely of lumber from his company's sawmills. From Weyerhaeuser's redbrick mansion overlooking the business district, clapboard-sided houses with peaked roofs and sturdy front porches stair-stepped down the hillside toward the Mississippi River. Near the riverbank, brick smokestacks rose from factories making plows, fertilizer, crackers, beer, and soap. Stern-wheeled steamboats threaded their way among the piers of iron-truss railroad bridges that led from Illinois to Iowa and then on to the West. On the far side of the river, the town of Davenport rose from the riverbank as if mirroring Rock Island in the flat gray sheen of the Mississippi.

Weyerhaeuser had not been spending much time in Rock Island lately. For the past few years he had been mostly in the forests of Wisconsin and Minnesota cruising timber or overseeing his crews. But somehow, whenever he returned home, everyone in Rock Island seemed to know about it, as if the *Argus* had carried the news in banner headlines. As his carriage wound down the hillside and into the commercial part of town, the townspeople turned to greet him, and he acknowledged them all with a nod of his head or a friendly wave. At forty-five,

Weyerhaeuser was a physically imposing man. Barrel-chested, with a full beard and mustache, he could still carry 120-pound sacks of wheat up a plank into a boxcar. Sometimes he joined his work crews in the forests or the mill, usually to show them the right way to do something. He maintained his health by drinking a quart of buttermilk daily and spending as much time outdoors as possible. He had a steady gaze and deep-set blue eyes that, according to a correspondent, "commenced to dance . . . when he was amused." Throughout his lifetime he was embarrassed by his German accent, often referring to Thursday as "the day after Wednesday" to avoid mispronouncing the *th*. He loved to tell stories and jokes, often to make a point that was apparent only in retrospect. People liked and trusted him. Up and down the Mississippi River from St. Paul to St. Louis, people said that no one was more honest and honorable than Frederick Weyerhaeuser.

In the spring of 1880, Rock Island—which just fifty years earlier had been the home territory of Black Hawk and several thousand members of the Sauk Indian tribe—was a bustling town of 20,000 people. The Chicago and Rock Island Railroad had reached the town in 1854, and since then Rock Island, which is situated on a bend in the Mississippi where the river runs west and then swings to the south around a large wooded island, had become the hub of a transportation network built of iron rails, freight trains, and steamboats. On the way to his mill, Weyerhaeuser went by stores selling baked foods, feed, harnesses, ice, toys, engravings, wallpaper, stationery, saddles, drugs, dried fruits, pork, and liquor. He passed the establishments of milliners, butchers, blacksmiths, locksmiths, farriers, barbers, and newspapermen. The adjoining city of Moline had been growing since John Deere established his manufacturing plant there in 1848, and Davenport had already grown larger than either city, but Rock Island considered itself the most established and consequential of the cities, and it resisted consolidation with its neighbors.

In all three cities, no commercial establishment was as impressive as the sawmills and lumberyards that Weyerhaeuser and his brother-in-law Frederick Denkmann had been building for the past twenty years. Their lumberyard, along with its associated mills and factories, stretched for nearly a mile along the eastern bank of the river. Stacks of lumber ready to be shipped to rapidly growing towns on the plains rose higher than adjoining buildings. Rafts of logs floating on the river were so dense that a person could walk halfway to Davenport without getting wet. Among the hundreds of sawmills scattered up and down the Mississippi River, Weyerhaeuser and Denkmann's was one of the largest and most profitable.

But as Weyerhaeuser passed through Rock Island that morning in 1880, he was thinking more about his problems than his accomplishments. For a decade, he and his associates had been engaged in a fierce competition with a group of lumbermen centered on Chippewa Falls in central Wisconsin. The issue was access to trees. By the 1870s the forests of Michigan, Illinois, and eastern Wisconsin were largely depleted. The lumbermen on the Mississippi needed more trees or their businesses would inevitably decline. Naturally, their attention turned to the rich forestlands of central and northern Wisconsin, including the watersheds drained by the Wisconsin, Black, Trempealeau, Buffalo, and Chippewa Rivers. In particular, the Chippewa, which originates near the Upper Peninsula of Michigan and empties into the Mississippi about sixty miles southeast of St. Paul, was a lumberman's paradise. Seventy miles wide and 160 miles long, the Chippewa drainage consisted of a broad plain of gentle slopes covered with stands of majestic white pine. As Weyerhaeuser told his son-in-law in 1901, "When I first saw the fine timber on the Chippewa, I wanted to say nothing about it. It was like the feeling of a man who has discovered a hidden treasure. If only Weyerhaeuser and Denkmann could control this, they would have an inexhaustible supply of the very best timber in the world."

But Weyerhaeuser was not alone in coveting the woods of the Chippewa River basin. A group of sawmill owners up and down the Chippewa had been logging nearby forests since the 1840s, and they were determined to keep the riches of the Chippewa to themselves. To newspapermen and their patrons in the Wisconsin legislature, they muttered about "aliens," "invaders," and "absentee capitalists" who were "seeking to wrest control of the government pine land from them." They delayed the downstream passage of logs purchased by the Mississippi

River lumbermen. On one occasion, when tensions were especially high, a crew of the Chippewa Falls millmen threw the manager of a competing sawmill into the river.

Weyerhaeuser knew that he could not win this battle by himself. He had always been most successful when he had spread the risks, when he had invited associates to help him with a new undertaking. In December of 1870, Weyerhaeuser and a group of Mississippi River sawmill owners had met in Chicago to form the Mississippi River Logging Company, a loose agglomeration of firms created to overcome the obstruction of the Chippewa Falls firms. The company's objective was to get the trees out of the woods and onto the Mississippi, where they could be rafted downriver to the separate sawmills of the partners. As Weyerhaeuser would later describe it, they were doing no more than the farmers in his hometown in Germany when they hired one boy to drive their cattle rather than each hiring a different driver.

The Mississippi River Logging Company set its sights on a side branch of the Chippewa called Beef Slough. A sluggish backwater that stretched for miles along the Chippewa where the river met the meandering channels of the Mississippi, Beef Slough was the perfect place to corral logs floated from upstream and sort them into rafts that then could be guided to the waiting sawmills downriver. But the Chippewa Falls lumbermen refused to yield. At first they resorted to cutting the booms and interfering with the log drives of the Mississippi River lumbermen. Then they moved the fight indoors. Several times, they unsuccessfully sued the Mississippi River lumbermen for interfering with the navigation of steamboats on the river. As one account summed up the situation, "By 1879 the Chippewa Basin was divided by a feud so bitter that the entire history of lumbering could hardly show its like. . . . The two groups glared at each other across an abyss as wide as the Grand Canyon."

The year 1880 was shaping up to be critical for the dispute. Overwhelmed by the management expertise that Weyerhaeuser and the Mississippi River Logging Company brought to their logging, the Chippewa Falls lumbermen were faltering. Already a large Chippewa Falls mill had gone bankrupt, and two-thirds of the logs coming down the

Chippewa belonged to the Mississippi River millmen. But Weyerhaeuser had no desire to crush the Chippewa Falls sawyers. On the contrary, he perceived a far greater opportunity. If he could merge the Chippewa Falls group with the Mississippi River lumbermen, he and his associates would control the logging for hundreds of miles along the Mississippi. Already he had begun offering olive branches to his adversaries. When water levels fell so low that a huge logjam formed on the Chippewa, he released enough water from a dam controlled by the Mississippi River lumbermen to free the jam. A month later, he delayed releasing water from a dam until the Wisconsin lumbermen had repaired a boom at Eau Claire. They were small gestures, but they meant much more.

In later years, some newspaper writers condemned Weyerhaeuser as a monopolist who sought to tie up the trees of the Midwest and artificially inflate prices. But Weyerhaeuser always had competition, both in the Midwest and later in the Pacific Northwest. The Mississippi River Logging Company had to compete with lumber from other parts of the country traveling on America's rapidly expanding rail network. Buying some logs and setting up a sawmill to convert that timber into lumber did not take much capital. And Weyerhaeuser never did anything alone. He always shared his ownership of companies with partners who were only too happy to defer to his preternatural business instincts. As he once said, "I have so many friends who want to make me rich."

• • •

If a novelist were searching for a model on which to base the story of a self-made nineteenth-century American tycoon, Frederick Weyerhaeuser would be an obvious choice. He was born on November 21, 1834, in the village of Nieder-Saulheim, Germany, in the rich rolling farmland southwest of Frankfurt. One of thirteen children sired by his father, he was the only boy to survive to adulthood, along with four sisters. His father, John Weyerhaeuser—the surname means "dweller in a house on a lake"—owned about fifteen acres of farmland, including three acres of grapes, and was much in demand for his ability to set out new vineyards. "Probably he wore himself out by over work," Weyerhaeuser later recalled, "for after an illness of nearly two years, beginning with a chill at his noonday lunch and developing into something like dropsy, he died Oct. 6, 1846, at the early age of fifty-two years."

Weyerhaeuser was eleven years old when his father died. As the only man in the house, he no longer had much time for school and soon gave it up to devote his full attention to the farm. He and his mother, with the assistance of hired men, were able to pay expenses and support the family. But they also began to feel the allure of America. Many farmers in the region were emigrating, attracted by reports of cheap lands and by what were known as "America letters"—correspondence from emigrants to America who wrote their European relatives about the rich, black loam of the prairies and the fortunes to be made in bustling new towns. By 1852 the imagined promise had become too great to resist. The Weyerhaeusers sold their farm and most of their belongings, packed what was left in steamer trunks, and departed for America. As Weyerhaeuser later reported of the trip from London to New York, heralding a toughness he would later demonstrate in the forests of the Midwest, "The voyage occupied about six weeks, and was rather rough at times; but I enjoyed every hour of it, and the harder the wind blew, the happier I was."

Following an older sister to northwestern Pennsylvania, Weyerhaeuser went to work for a brewer, getting four dollars a month the first year and nine dollars a month the second. Soon he was running the brewery, but he soured on brewing when he saw that "a brewer was often his own best customer." He thought of buying a farm south of Erie with his portion of the proceeds from selling the farm in Germany, but the Pennsylvania farmhouse struck him as such a lonely place that he decided against it. One of his father's cousins had moved to a farm south of Rock Island and had sent his relatives effusive accounts of life near the Mississippi. Weyerhaeuser traveled to Illinois in March 1856 to visit his relative in the countryside, but farming, once again, struck him as a dead end, and he settled instead amid the bustle of Rock Island. For a few weeks he carried chain for the surveyors laying out railroad lines in Rock Island. He briefly returned to work as a brewer but was on the lookout for something better.

A friend of his was working as a fireman—the worker who tended the steam boilers—at a small sawmill on the river, and he talked Weyerhaeuser into taking the night fireman job. Two days later the mill owners canceled the night shift, but they kept Weyerhaeuser on, putting him to work tallying the amount and quality of the lumber emerging from the mill. He rose quickly through the ranks. The owners were impressed by his intelligence, friendliness, and business smarts. Soon he was in charge of all sales in the lumberyard; before long he was looking after the entire business. "The secret of this lay simply in my readiness to work," he later said. "I never counted the hours or knocked off until I had finished what I had in hand."

Back in Erie he had met Sarah Elizabeth Bloedel, who also had immigrated from Nieder-Saulheim. In the spring of 1857, when she was eighteen, Sarah came to Rock Island, ostensibly to help her older sister, Mrs. Frederick C. A. Denkmann, through a difficult childbirth, but also, most likely, at the urging of the young countryman she had met in Pennsylvania. Six months later they were married. They moved shortly thereafter to the nearby town of Coal Valley, a few miles south of Rock Island, so that Weyerhaeuser could run one of his employer's sawmills. Within a year, children started arriving with a regularity that one would later call "quite appalling": John in 1858, Elise in 1860, Margaret in 1862, Apollonia in 1864, Charles in 1866, Rudolph in 1868, and Frederick E. in 1872. People would later remember the Weyerhaeusers sitting in their small home in Coal Valley, doors and windows wide open, Sarah sewing and Frederick working at his ledgers. Weyerhaeuser later told his eldest son, "The happiest days of my life were in Coal Valley."

Shortly after the Weyerhaeusers moved to Coal Valley, his employers bought a raft of logs from an unscrupulous raftsman who never delivered, and the company soon went bankrupt. A creditor of the company asked Weyerhaeuser if he wanted to buy the equipment in the mill and keep it running, which he did on credit. But he could not afford to buy the mill itself, so he approached his brother-in-law, F. C. A. Denkmann, with a proposal. If Weyerhaeuser put up most of the cash that he had made from running the mill, would his brother-in-law pitch in what he could? Denkmann agreed, though the final stake was a strange assortment of assets, including two buggies, a wagon, four harnesses, three horses, four cows, seven hogs, one log chain, two saws, and a skiff—virtually everything the two men owned except for their houses and furniture. Thus was born, on May 1, 1860, the firm of Weyerhaeuser and Denkmann, names that soon would be known throughout the Midwest.

At that time, sawmill owners in the region had two basic ways of getting logs for their mills. They could buy them from suppliers, generally as rafts of logs floated down the Mississippi River from the forests of Wisconsin. Or they could buy property (or the right to cut down the logs on property owned by someone else, which is known as stumpage) and then hire logging teams to harvest the logs and deliver them to the mills. Weyerhaeuser and Denkmann began by taking the former approach. Denkmann, a high-strung but shrewd man who was also an excellent mechanic, kept the equipment running at the mill. Once, when his hand got caught in the planer knives, he lost several fingers, but he was back at work the next day. Weyerhaeuser dealt with suppliers, creditors, and customers. He had a superb mind for business. He could instantly assess the tangled threads of a problem and begin working on solutions. He was a tough bargainer, but so honest that immigrant farmers who could not read his bills paid him without question. He paid his employees on time, in full, and usually in cash.

It was the perfect time and place in American history to be running a sawmill. When the Civil War ended in 1865, many demobilized soldiers were paid in scrip, which they could exchange for property. This added momentum to a land rush that was rapidly spilling beyond the Mississippi onto the Great Plains. The settlers needed wood for homes, barns, fences, tools, and fuel. They were completely dependent on lumber cut from distant forests and sold by companies like Weyerhaeuser and Denkmann; even sod homes needed timber bracing. Meanwhile, the rapidly building cities of the Midwest needed wood for new construction and for rebuilding after fires. When, later in the century, paper began to be made from wood pulp rather than from cotton rags, the importance of wood to the American economy became even greater.

Another major source of demand for lumber was the same institution that was opening the prairies for settlement—the railroads. The completion of the first transcontinental line in 1869 had set off a mad scramble to extend more lines into the West. By the 1880s the nation's railroads rested on nearly 400 million eight-foot-long-by-eight-inch-square ties. Wood was needed for bridges, station houses, platforms, and fuel (until coal began taking over in the 1870s). In the nineteenth century, railroads were sometimes known as the iron road, but they used far more wood than metal.

Weyerhaeuser was lucky in another way. Before the Civil War, manufacturing enterprises were generally too small to generate large revenues, so manufacturers had to turn to merchants and bankers for capital. But in the latter half of the nineteenth century, manufacturing facilities like Weyerhaeuser and Denkmann's sawmill became large enough to provide capital for further expansion. Manufacturing became a source of great wealth, especially when tied to a plentiful natural resource. The Gilded Age, from which sprang the Carnegies, Rockefellers, and Weyerhaeusers, was built on a foundation of iron, oil, and wood.

At first the firm of Weyerhaeuser and Denkmann bought logs from suppliers who rafted the wood down the Mississippi to the mill. But as the firm grew, Weyerhaeuser needed more and cheaper logs, and the next step in the expansion of the business was an obvious one. In 1867 Weyerhaeuser made his first purchase of stumpage. Two years later, he bought several stands of timber along the Chippewa River. It was the beginning of what would become a woodlands empire.

Buying stumpage, and then timberland, inevitably took Weyerhaeuser away from his rapidly growing family in Rock Island. He began spending weeks in the woods of Wisconsin, and then months, and then entire winters. He missed his wife and children dearly, but he also found peace in the woods of Wisconsin and Minnesota. "I love the woods life," he once told an associate, and he never complained of the privations he suffered there. He slept in backwoods camps with his crews, all the men laid out next to one another under a single blanket. He helped his crews haul fallen logs through fierce snowstorms to frozen riverbanks. Once, when a protruding mole on his face was beginning to detract from his appearance, Sarah was pleased to discover, on one of his trips back to Rock Island, that it had frozen off during a particularly harsh winter.

Weyerhaeuser, always a quick study, learned things in the woods that he never would have learned in an office on the banks of the Mississippi. Soon he could tell at a glance how many board feet a skilled sawyer could get from a standing tree. He learned what made workingmen feel content and what caused them to burn with resentment. He rode his own log rafts down the Mississippi, watching the bluffs and islands slip past as if the only stationary objects were the river and the logs on which he stood. He learned to apply what he called "intelligent courage" to his business so that he could act when other men would have hesitated. He grew strong and healthy in the north woods, and as his business grew, so did his wealth.

Weyerhaeuser was as religious as any of his countrymen. He went to church whenever he was in Rock Island and had a special suit reserved for Sundays. But an episode from his youth in Germany had made a powerful impression on him. When the old stone church in his village was torn down, coffins that had been buried in the basement were exhumed and opened. At first the bodies inside seemed miraculously preserved. But when they were exposed to the air, they crumbled into dust. It is hard to know what Weyerhaeuser made of this story. But the fact that he remembered it so vividly suggests that he was more likely to place his faith in this world than the next one.

• • •

Almost by accident, Weyerhaeuser had joined a uniquely American enterprise. With the exception of agriculture, no industry has had a greater influence on the physical appearance, the economic prosperity, and the moral economy of America than logging. Wood propelled the European colonies in the New World, and then the young United States, to global prominence. It sustained generations of settlers, farmers, and industrialists who found on the frontiers of civilization a resource of unparalleled versatility. It conditioned how Americans thought about their country and its future. As one account of America's forests has put it, Americans are, "above all else, products of the continent's forested plenty."

When the first Europeans arrived on the shores of North America, they were astonished by what they saw. In Europe, most of the land had been cleared and planted with crops or forage for centuries or millennia. But in North America, a great green blanket of trees rose at the water's edge and extended as far as the eye could see. Though Indians had cleared and reworked parts of the forests, trees covered most of the eastern half of North America. In what was to become New England stood immense stands of white pine, spruce, maple, and birch. Throughout the Mid-Atlantic and Midwest were maple, oak, birch, chestnut, and hemlock. The South had yellow pine, tupelo, cypress, and sweet gum. It was the greatest temperate-zone forest on earth.

Mortality was high among the European colonists, but far more would have died without the forests of the New World. The colonists hunted rabbit, fowl, and deer in the woods. They ate cherries, acorns, chestnuts, and walnuts that hung from trees free for the taking. They chopped down the forests for firewood, stockades, and logs with which to build cabins. They converted maple sap into sugar and extracted medicines from roots and bark. So beneficent were the forests and croplands of North America that the colonists and their descendants soon had a much higher standard of nutrition and health than did their counterparts in Europe.

But the woods were also threatening to the early colonists. They sheltered bear and cougar, Indians, and perhaps, in the minds of the excessively religious, Satan himself. Many settlers found the never-ending forests gloomy, oppressive, and overbearing. The woods were a barrier to roads, fields, townships, and landscapes such as they had known back in Europe. As one wrote of Pennsylvania in 1681, "It is a very fine Country, if it were not so overgrown with Woods."

They attacked the forests with grim resolve. They girdled trees by removing a strip of bark from around the base of the trunk, which cuts off the flow of nutrients from the leaves to the roots and kills the tree. Like the Indians before them, they set fire to forests to open the land for agriculture. They chopped down trees, split logs for firewood, and pulled out the stumps with teams of oxen. To many colonists, a tree was no more than a challenge to the ax.

In a land of wood, the colonists also began to experiment with this ever-present material. Many items made of stone, metal, leather, or clay in Europe were fashioned from wood in America. The settlers of the eastern seaboard used wood to make utensils, bowls, boxes, toys, cradles, matchsticks, barrels, furniture, tools, clapboards, shingles, paneling, windmills, musical instruments, coffins, bridges, fences, and wagons. They built their houses, outbuildings, and even roads with wood, and when the wood rotted they simply tore out the old structures and built new ones. As one traveler from Europe reported to his countrymen, visitors to America should be prepared for "a Wooden Town in a Wooden Country & a Wooden bred set of Tavern-keepers."

The colonists also turned to wood as an export to sustain their economy. At first they shipped both logs and roughly cut boards, planks, staves, and shingles to Europe for manufactured goods. Particularly valuable were masts, spars, and bowsprits from the pineries of New England for the ships of the Royal Navy, which traveled the world using sails hung from American trees. Soon Americans developed industries of their own based on cheap and plentiful wood. Shipbuilding was New England's most profitable export business during colonial times. Tar, pitch, and turpentine extracted from trees—collectively known as naval stores—supplied both American and European shipbuilders.

Sawmills for local construction and exports to Europe spread along the eastern seaboard and into the hinterland along riverways. At the mouths of those rivers grew towns constructed of wood cut upriver and floated to consumers: New York, Philadelphia, Baltimore, Richmond. Along the East Coast, sawmills, towns, and then cities sprang up at the fall line, where the last waterfalls before tidewater provided power for the saws. (Today, Interstate 95 roughly follows the fall line up the East Coast.) By the end of the 1600s, more than seventy sawmills were active in the Massachusetts Bay Colony alone.

Nowhere was the timber industry more important to the economy than in Maine. Wealthy merchants and lumbermen in the Massachusetts Bay Colony, looking for places to invest their profits, purchased immense tracts of timber and built sawmills where rivers met the sea. The industry worked its way northeast, tapping into the forests of the Piscataqua, Saco, Presumpscot, Androscoggin, Kennebec, Penobscot, Machias, and St. Croix river valleys. By the time Maine became a state in 1820, it was the leading logging region in the country. The towns of Bath, Bangor, and Machias were among those that grew wealthy serving the lumber industry. As Henry David Thoreau wrote in his book *The Maine Woods*, the mission of the men on the Penobscot "seems to be, like so many busy demons, to drive the forest all out of the country, from every solitary beaver-swamp and mountain-side, as soon as possible."

No New England tree was more prized than white pine. As much as 250 feet tall and 3 feet wide at the height of a man, a tall, straight white pine could have 100 feet of clear, unknotted wood before the first limb appeared above the forest floor. It resisted shrinking, splintering, warping, or cracking. It held paint and varnish. It gripped nails firmly, even when a nail was removed and then driven back into the same hole. It was light, strong, durable, and easy to carve. The figureheads of whalers, clippers, and frigates throughout the world began their existence as old-growth white pines in the forests of New England.

Maine is also where popular conceptions of lumberjacks in the United States first took hold. As men flooded into the woods to cut down the forests of New England, a mythology sprung up that the men, consciously or unconsciously, sought to uphold. They were rough, boastful, crude, boisterous—the opposite of the laconic cowboy. They argued, lied, gambled, and fought. They tried to outdo one another both in the woods and in the bunkhouses, where their tall tales contributed eventually to the creation of Paul Bunyan and his blue ox, Babe. As Timothy Dwight, president of Yale from 1795 to 1817, wrote:

> These men cannot live in regular society. They are too idle, too talkative, too passionate, too prodigal, and too shiftless to acquire either property or character. They are impatient of the restraints of law, religion, and morality; grumble about the taxes by which rulers, ministers, and schoolmasters are supported; and complain incessantly, as well as bitterly, of the extortions of mechan-

ics, farmers, merchants, and physicians to whom they are always indebted.

The lumberjacks did tedious, backbreaking, and perilous work. They lived in camps, where they slept in common beds, ate in dining halls, and spent their free time in crowded common rooms. The most commonly reported feature of the camps were the bugs, the snoring, and the stench—an indescribable miasma composed of wet wool, unwashed men, tobacco juice, and whiskey (when the loggers could get it). They wore brightly colored plaid shirts to be seen in the woods, canvas pants soaked in paraffin to keep out the rain, fur or wool caps to stay warm, and boots with nails driven through the soles for traction on slippery logs—known as calked boots (though the word has always been pronounced "corked").

With the possible exception of whaling, no job was more dangerous for the men of New England. A miscalculation by a faller could send a ten-ton tree headed straight toward a knot of men. The logs piled in a landing could suddenly collapse, crushing workers under an avalanche of wood. Men were maimed by axes, trampled by panicked oxen, blinded by wood chips, and crippled by logs rolling down hillsides. When a man was killed, his colleagues hung his boots on a nearby tree or streamside bush as a memorial for his compatriots. Logging made men strong or it killed them, and the ones who survived were proud to have cheated death.

The loggers of New England developed the techniques of white-water logging that later were emulated across the country. In winter, when the ground was frozen solid and covered with snow, they cut trees and dragged them with oxen to landings next to frozen streams, where the logs were jimmied into great piles. Each log had a brand chopped or seared into its cut end identifying the company that owned it. Come spring, the lumberjacks pried the logs into the roiling streams to be carried to the sea. Sometimes, when the spring runoff was low or the harvest was high, the logs would catch on ledges and sandbars and form immense logjams that could extend for miles along a river. One logjam on the Kennebec River contained an estimated 25,000 logs stacked as much as ten feet high. Then workers known as river pigs would climb onto the jam with their peaveys—long poles with a metal spike extending from the end—pry at the logs until they suddenly sprang free, and try not to be crushed in the ensuing melee of timber, water, and scrambling men.

The loggers of Maine and their successors across the United States have extracted an immense amount of wealth from the forests of North America, though very few ever got wealthy themselves. The exact value of the wood taken from America's forests is impossible to calculate, but rough estimates can be made. When the colonists landed along the Atlantic Coast, about half of what would become the contiguous United States was forested. Today about 40 percent is. But a more telling statistic is the amount of timber harvested from America's forests. In 1620, what would become the contiguous United States had an estimated 7.5 trillion board feet of standing sawtimber—with a board foot of wood equal to a board 1 inch thick and a foot square. Today the number is about 2.5 trillion board feet. Americans have reduced the volume of the wood in their forests by about two-thirds, or 5 trillion board feet, since the Pilgrims landed in Plymouth.

Over that time, new wood has grown in America's forests, some of which still stands, some of which has burned or rotted, and some of which has been harvested for lumber or firewood. Based on numbers from the US Forest Service, the amount of new wood harvested from America's forests since the arrival of European settlers can be pegged, again very roughly, at about another 5 trillion board feet. Of course, only part of America's primordial and newly grown forests have been converted into lumber. But let's say that half of the existing and newly grown wood in the United States has ended up as lumber (and, in the twentieth century, as other wood products such as plywood, veneer, and pulp)—a number that approximately matches separate estimates of the amount of wood products used in the United States over its history.

At today's retail prices, the value of the wood products extracted from America's forests would be about $2.5 trillion dollars. That's at least five times the value of all the gold extracted from the United States since its founding.

• • •

In June of 1880, Weyerhaeuser's battle with the Chippewa Falls lumbermen was coming to a head. He and his associates had to gain control of the Wisconsin forests or their businesses were doomed.

That month, it began to rain in the upper Midwest, and the rain did not stop. The waters of the Chippewa River began to rise. First the water rose to the edge of the riverbanks. Then it crept over the riverbanks and lifted the logs left to raft to the mills. The logs entered the swift-moving current, forming a loosely connected projectile of immense speed and weight. This projectile acted as a battering ram against the piers, bridges, and structures along the river. The rising water carried away barns, homes, and even an entire sawmill. Two bridges at Chippewa Falls failed, adding to the debris in the river. Finally, as the *Pioneer Press* reported on June 14, 1880, "The disaster on the Chippewa river expected for two or three days occurred yesterday morning, when the great Eau Claire boom gave way and started two hundred million feet of logs on their journey to the Mississippi."

When the flood finally subsided, more than half the logs belonging to the Chippewa Falls lumbermen were below their sawmills, with no possible way of getting them back upriver. The only mills that could cut those logs now were those on the Mississippi River. In the battle with their downstream competitors, the Chippewa Falls lumbermen had often blocked the passage of logs belonging to the Mississippi River sawmill owners. Now they expected retribution and almost certain ruin.

Weyerhaeuser had a different idea. Even before the floodwaters receded, he dispatched three steamboats and sixty men to round up the logs that had been swept downstream. Then he called a meeting of the lumbermen on both the Mississippi and Chippewa Rivers. He proposed that for every Chippewa Falls log cut by the Mississippi River lumbermen, the Chippewa Falls lumbermen would receive a log owned by the Mississippi River millmen in exchange. In the end, the amount of lumber produced that year exceeded the previous year's total.

With that act of generosity and cooperation, the opposition of the

Chippewa Falls lumbermen to the Mississippi River group collapsed. In November 1880, at the Grand Pacific Hotel in Chicago, the Mississippi River and Chippewa Falls lumbermen agreed to work together to purchase, drive, and distribute logs to the group as a whole. Weyerhaeuser received permission to buy properties for the group as he saw fit. As one associate said of him, "No other man in America knows so much about pine as he does."

Weyerhaeuser was at the head of the largest logging conglomerate ever seen in America. But he was just beginning to build his empire.

THE PURCHASE

IN 1891, WEYERHAEUSER MOVED HIS FAMILY FROM ROCK ISLAND TO St. Paul to be nearer his new base of operations in the pinewoods of Minnesota. In St. Paul he bought the mansion next to railroad baron Jim Hill's on Summit Avenue. Partly his new house must have reminded him of his home in Rock Island. From his backyard, one- and two-story houses trailed down the hillside toward the Mississippi River. Smoke from hundreds of downtown factories formed a sweet gray-brown haze that filled the valley. From the riverbanks could be heard the whistles of distant trains, probably from Jim Hill's locomotives.

Weyerhaeuser also must have been intrigued by his next-door neighbor. Jim Hill was not only the most powerful man in all the land west of Chicago, but his experiences were so like Weyerhaeuser's as to be uncanny. They quickly became friends and spent many evenings together in the drawing rooms of their houses. Their only difference was that, after many years in the woods, Weyerhaeuser fell asleep early in the evening, while Hill had endless stores of energy and seemed never to sleep. Their families often laughed when they passed the door to the drawing room and saw Hill soliloquizing while Weyerhaeuser sat fast asleep in an armchair.

In the fall of 1899, Weyerhaeuser had good reason to stay awake.

Because of his railroad dealings, Hill had gained control over one of the most tantalizing forests left in the United States. As Hill had said of his possession, "The timber between the Cascade Range and Puget Sound is the largest and of the best quality I have ever seen. The amount of merchantable lumber per acre is from six to ten times as great as the best Michigan or Wisconsin timberland. It is impossible to realize the immense growth of these trees without seeing them."

Weyerhaeuser knew that Hill wasn't exaggerating. He had seen those trees himself—on a trip to the Northwest in 1891. It was the most magnificent forest imaginable. The trees clung thickly to the steep hillsides, rising in an immense expanse of green from the water's edge to distant glaciated mountains that shimmered in the midday sun. It was more wood than human beings could ever use, a limitless supply, just waiting for someone to begin gnawing at its margins. And on this summer day in 1899, Weyerhaeuser needed those trees. People had said the same thing about the forests of the Midwest—that they could never be conquered, that the wood would last forever. Yet he could see that the trees were thinning. The best trees were already gone from many forests; soon the loggers would take the rest. The forests would be left as cutover and desolate as they were in Michigan, and in New York and Pennsylvania before that, and in Maine before that.

Jim Hill had more trees than any man in the United States had ever owned. Weyerhaeuser had money, and he knew that railroad men always needed money. The question was how best to exchange one for the other.

Hill was born four years later than Weyerhaeuser, on September 16, 1838, on a hardscrabble farm near Guelph in the Canadian province of Ontario. The elder of two boys, he spent much of his childhood hunting and fishing with his brother. When he was nine, a bow he was making for his brother snapped, propelling an arrow back into his right eye socket. An ingenious backwoods doctor got the eye back into place and restored muscular function. But the eye could never see much more than shades of gray. It never bothered Hill. His one good eye served him well during a long and exceedingly productive life.

When Hill was fourteen his father died, and shortly thereafter he quit school and went to work keeping the books and doing chores at a general store in the small town of Rockwood. Like Weyerhaeuser, he was a quick study. He gained more and more responsibility as he learned about supply chains, distribution networks, price negotiations, and the vagaries of agricultural commerce. But Hill was too ambitious and restless to stay in a small town in Canada forever. An ardent admirer of Napoleon, he took the middle name of Jerome, after the emperor's brother, at age thirteen. James J. Hill had his one good eye set on the world.

In 1856, at the age of seventeen, Hill moved to St. Paul. It was the perfect place for a budding entrepreneur. Less than two decades old, St. Paul already had more than 10,000 people, easily eclipsing the small town of Minneapolis a few miles farther up the Mississippi River. As the head of navigation on the Mississippi, St. Paul was the logical center of transportation and commerce in the upper Midwest. People and goods got off steamboats and boarded railroads and wagons for the journey west, while goods from the interior flowed in the opposite direction to the growing cities of the Midwest.

Hill soon landed a job as a shipping clerk for a fleet of Mississippi steamboats. Strong, gregarious, independent, and an incredibly hard worker—he was known for working late, going for a swim in the river, downing a cup of strong coffee, and going back to work—he fit right in on the bustling river levees. He saved his meager but growing earnings and soon invested in a small warehouse on the river. A few years later, he established his own shipping business, which required that he work closely with the fledgling St. Paul and Pacific Railroad. Despite its grandiose name, the St. Paul and Pacific extended no farther than the rich farmland of the Red River valley on the border between Minnesota and Dakota Territory. It was an undercapitalized line that its financiers had used mostly to skim profits from overpriced construction contracts. Minnesotans referred to it as "two streaks of rust and a right of way." Yet Hill was fascinated by the decrepit, unfinished, and soon-to-be-bankrupt line. If he could gain control of the St. Paul and Pacific, he would have an asset that no one else had: rail access to a part of the country that might be backward and isolated for the time being but would not stay that way.

In 1877, with three and later four partners, Hill made a lowball offer for the railroad, and in early 1878 its owners agreed to sell. By the early 1880s, Hill had essentially given up his other businesses to turn to the task for which he seems to have been born: railroad builder. The reorganized and renamed St. Paul, Minneapolis, and Manitoba Railway Company soon became a thriving line that made Hill a multimillionaire. He knew every detail of the railroad's operations—down to the names of the engineers who drove the trains—while at the same time he provided the broad vision that gave the line momentum and purpose. Between 1879 and 1883 the railroad line added more than nine hundred miles of new track. Hill traveled the tracks obsessively, looking for any way to straighten curves and reduce grades. He replaced the old iron rails with tempered steel, often stopping to talk with the gangs of tracklayers and graders working on the line. He built a massive new headquarters building in St. Paul, terminals in the cities and towns served by the railroad, and hotels to cater to wealthy tourists—all while fathering ten children with his wife, Mary, and building a massive Richardsonian Romanesque mansion on Summit Avenue that remains one of the great tourist attractions of St. Paul today.

Hill was a relative latecomer to the railroading frenzy of the nineteenth century. By 1885, four separate transcontinental lines spanned the nation. In general, these lines were poorly built and underused. The railroad barons of the nineteenth century were more interested in lining their pockets than in constructing a solid transportation system. The railroads went bankrupt repeatedly in the decades after they were constructed even as their financiers reaped millions. Hill knew, from his experiences with the Manitoba, that great riches were still available to someone who could build a heavy-duty, high-volume railroad that could create new markets, not just serve existing ones. At the age of fifty, he was about to embark upon the greatest adventure of his life.

The building of the Great Northern Railway is a romantic story in an industry given to romanticism. By 1886, the Manitoba had laid track west from the Red River to Minot in Dakota Territory, a town named after one of Hill's vice-presidents. Now came the turning point. Could

Hill transform a small but solid regional rail line into a fifth transcontinental railroad? The Pacific Ocean was a thousand miles away, and those thousand miles were a barren wasteland of plains and high mountains with very few settlers, towns, or businesses along the way. Building a railroad that was designed to be profitable and not just a moneymaking scheme for investors was an audacious leap of faith. But Hill had faith in abundance.

The first push was to Havre, Montana, where spur lines led south to the rich mining regions around Helena and Butte. For a while Hill considered trying to gain control of the northernmost transcontinental, the Northern Pacific Railroad, rather than building his own line west from Great Falls. But in 1889, even as Hill renamed his railroad the Great Northern Railway after a line he admired in Great Britain, he chose instead to push directly west over the high Montana Rockies to the area of Flathead Lake in Montana, to Spokane Falls in eastern Washington State, and on to what was then the small lumbering and fishing town of Seattle. Aided immeasurably by the discovery of a low pass over the Rockies just south of what is today Glacier National Park, crews worked feverishly to build lines through narrow canyons, over high wooden trestles, down heavily wooded valleys, and across some of the most isolated and beautiful country in the nation. They made steady progress. As Hill is reported to have said, "Give me enough Swedes and whisky and I'll build a railroad through Hell."

In June 1893, the first passenger train from St. Paul took four days to reach downtown Seattle, where, on July 4, a great celebration was held to commemorate the end of Seattle's isolation from the rest of the nation. The line opened just in time for passengers to travel to the World's Fair in Chicago—for $35 in first class and $25 in second class.

• • •

The next month, the Panic of 1893 began when the precarious financing of railroads caused a series of banks to fail. So many people walked away from recently constructed Victorian houses and left them empty that the legend of the haunted Victorian took root. L. Frank Baum wrote a story about a girl from Kansas and a wizard that has many parallels with the economic and political events of the decade. Most important, the Northern Pacific, Hill's main competitor in the northern tier of the United States, went bankrupt yet again.

The NP's bankruptcy presented Hill with the greatest opportunity of his life. Together with the financier J. P. Morgan, Hill moved to gain control of the line and merge it with the Great Northern. The result would be the greatest rail line ever known in the United States—an industrial powerhouse that would dominate the northern half of the United States from Chicago to Seattle. But this was too much for a federal government that was finally beginning to challenge the Gilded Age's growing concentration of economic power. In 1896 the Supreme Court barred Hill and his associates from merging the two lines. (Only in 1970 did the Supreme Court relent, allowing the merger of the Great Northern and Northern Pacific, which resulted in the creation of the Burlington Northern railroad.) Nevertheless, Hill and Morgan succeeded in establishing an alliance between the two, with Hill at the helm. And in the process, Hill gained control of the single most attractive feature of the Northern Pacific: its incredible land grants.

Historians have been arguing almost since the day these railroad land grants were made whether they were a necessary spur to western development or the greatest transfer of public wealth into the pockets of conniving plutocrats in US history. (Actually, they were both.) From today's perspective, the idea is absurd. In the 1860s and '70s, the US Congress passed legislation that ended up granting ownership of about 130 million acres to the builders of various railroads, with another 50 million acres coming from state land grants. That's an area four times the size of New England, a quarter again larger than California, and a bit less than 10 percent of the contiguous United States. The history of the West cannot be separated from the influence of the land grants on the ownership, development, and use of western lands.

No land grant was bigger than the one given to the Northern Pacific. The company eventually claimed almost 40 million acres of land—an area greater than the nation's nine smallest states combined. Under legislation signed by President Lincoln in 1864, Congress gave the railroad twenty square miles of land for each mile of track laid in states and forty square miles of land for each mile of track laid in the territories. The land was granted in a bizarre checkerboard pattern still visible on many western maps today. If the land really were a checkerboard, the railroad would acquire the white squares while the government retained the black squares. The idea was that the presence of the railroad would increase the value of the government's land. The railroad could sell its land to finance the construction of the line, and the federal government would benefit by the increased value of the land it retained. Everyone would come out ahead.

The legislation meant that the railroad's property extended for twenty miles on either side of the tracks in states and for forty miles on either side in the territories. The summit of Mount St. Helens was thirty-five miles from the Northern Pacific line built from the Columbia River to Tacoma in the 1870s, when Washington was still a territory. That's why, when the mountain began to shake in 1980, the top of the volcano was owned by the Burlington Northern Railroad.

Yet, in a bizarre historical footnote, the Northern Pacific had nearly lost its land grant, including the land around Mount St. Helens, long before James Hill became involved with the railroad. Shortly after Congress passed the original Northern Pacific land-grant legislation in 1864, it added a stipulation that the Northern Pacific must have a line completed to Puget Sound by the end of 1873 or it would lose its land grant. But this qualifier did not specify where the tracks had to start—only that they reach Puget Sound. Thus, in 1870 the Northern Pacific began laying tracks northward from Kalama, Washington—a small town north of Portland where the Columbia River makes its final bend westward toward the sea—bound for Puget Sound a hundred miles to the north. The tracks extended from the Columbia River north along the Cowlitz River, over a rise of land south of the present-day town of Chehalis, and through a series of broad drainages to the town of Tacoma, which in 1870 was a hamlet of one hundred people, including a sawmill, a public school, a hotel, and a store.

By June 1873, the tracks had reached Tenino, about forty miles south of Tacoma, and the project was on schedule. But in that month Jay Cooke, the financier who had been bankrolling the construction of the

Northern Pacific, went broke, which sparked a nationwide financial meltdown. Money for wages ran out, and workers began quitting. At one point, unpaid workers seized a bridge near Tacoma demanding $73,000 in back pay. As the situation grew dire, the Northern Pacific began selling land near Tacoma to pay the tracklayers and save its land grant. The crews struggled on. A few days before the deadline, the first work train eased onto the newly laid tracks next to Tacoma's Commencement Bay—and promptly pitched into the mud when the tracks gave way. No one seemed to care. Tacoma had achieved the supreme objective of every well-placed and ambitious western port town—to be the terminus of a transcontinental railroad.

Today the rail line from Portland to Seattle seems like a poor neighbor to the massive freeway that roughly parallels its course. Yet the railroad tracks in the shadow of Interstate 5 represent a critical turning point in Washington State history, because they delivered into Jim Hill's hands some of the most valuable land existing anywhere on the planet.

• • •

In Jim Hill's dining room in St. Paul, Weyerhaeuser faced the toughest choice of his life. The woods in Wisconsin and Minneapolis were headed toward extinction. He could call it quits and begin shutting down his company, as many of his associates on the Mississippi River were doing. He had set up his sons in the lumber business in Minnesota, and they were doing well. They were smart boys, Yale educated, as Jim Hill's sons were. They would do fine for themselves. But there was no long-term future in the cutover river basins of the Midwest. If he wanted to do something really useful for his sons, and for their sons, he would have to be bold once again. And as Weyerhaeuser often told his associates, "the only times [I] ever lost money on timberlands were the times when [I] didn't buy."

On January 3, 1900, Weyerhaeuser and Hill announced one of the largest private land purchases in US history. For $6 an acre, Weyerhaeuser and his business partners would buy 900,000 acres of timberland in southwestern Washington State. Eventually, word leaked out about the negotiations in the Summit Avenue houses. Weyerhaeuser offered $5 an acre, Hill demanded $7, and they split the difference.

At the time, the purchase was highly speculative. Railroad foresters had cruised some of the Northern Pacific timber to gauge its value, but the rest of it was bought sight unseen. Weyerhaeuser and his associates had no way to harvest or saw all that timber, and wildfires could at any time burn their investment to the ground. Many of the partners who provided part of the $5.4 million purchase price were skeptical. They thought they had been taken, especially when rumors surfaced that the Northern Pacific was selling other land for much less. But when some of Weyerhaeuser's partners started talking about giving up the land by not paying taxes, he suggested that they deed the abandoned land to him instead, which quickly ended the discussion.

In fact, the purchase was one of the best deals that anyone has ever made anywhere. Though the quality of the land varied, Weyerhaeuser and his associates had just bought some of the richest forests on earth. Many tracts had more than 100,000 board feet of timber per acre—breathtaking old Douglas fir and red cedar trees that had been growing since the Middle Ages. Their trunks were so huge that entirely new logging methods had to be developed to harvest them. As Weyerhaeuser wrote modestly of his newly purchased timber, "There is a great lot of it in every conceivable direction."

Any estimate of the value of that wood is of course tentative. But a ballpark figure, after correcting for inflation, is that Weyerhaeuser and his associates earned a return of more than $250 for every dollar they spent on Jim Hill's land in 1900. Furthermore, the company continued to buy land at about the same price for the next few years, until by 1905 it owned a million and a half acres of Pacific Northwest forestland. Of course, the company would have to build facilities and pay men to cut, saw, and sell all that wood. But everything left over would be the sweet, sweet fruit of capitalism.

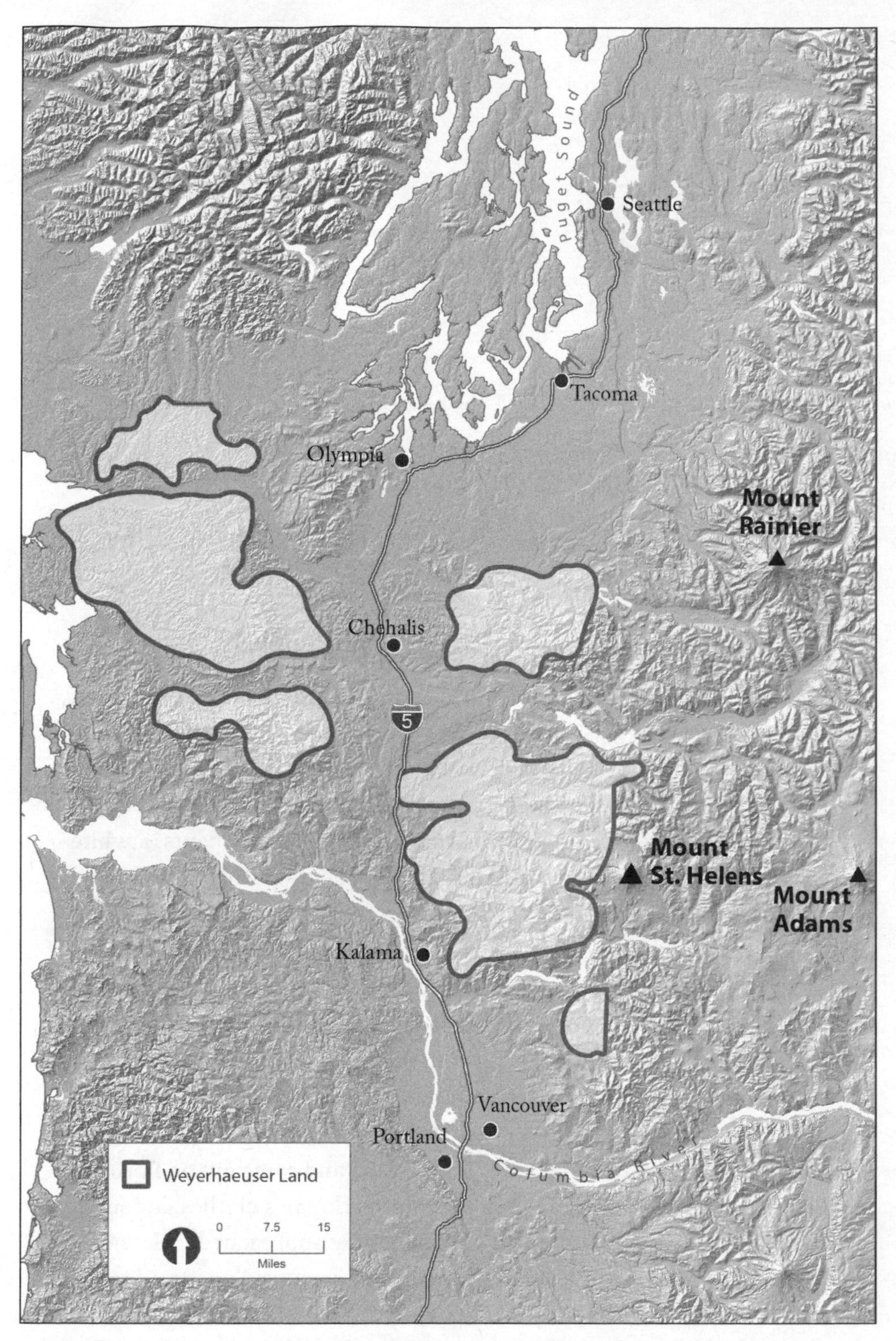

Land owned predominantly by Weyerhaeuser in southwestern Washington State

THE ABDUCTION

GREAT WEALTH HAS ITS PRIVILEGES—AND ITS PERILS.

On May 24, 1935, the great-grandson of Frederick Weyerhaeuser, nine-year-old George Weyerhaeuser—who in 1980 would be head of the Weyerhaeuser Company when Mount St. Helens began to erupt—was walking home from school along the broad and leafy streets of north Tacoma, Washington. He was wearing brown corduroy pants, a white sweatshirt, and white tennis shoes, and he occasionally jumped as he walked, because it was a Friday afternoon, because he was planning on trying out for the track team the next year at school, and because it was a sunny spring day, almost too warm for the sweater he was wearing on top of his sweatshirt. Normally he would have waited at his sister's private school for the chauffeur to bring them home, but George's public school fifth-grade class had let out fifteen minutes early, and, on a whim, he had decided not to wait for her. He cut through an overgrown path that bordered the Tacoma Lawn Tennis Club and came out on Borough Road. A nervous-looking man dressed in workman's clothes got out of the passenger seat of a running car as the boy approached. "How do I get to Stadium Way?" the man asked. George shrugged and tried to get past him. But the man grabbed the boy by the shoulders, led him to the car, pushed him onto the floor of the backseat, and covered him with a blanket. Then the man got into the passenger seat of the car, and the driver sped away.

They drove a long way, with George covered by the blanket. Finally the car stopped and George heard the front door open. "Get out," the man ordered. When George emerged from the car, he saw that they were in a sea of stumps, a cutover forest east of Seattle. The men were wearing hoods with rough holes cut for their eyes. They had George sign the back of a piece of paper. Then they blindfolded him and led him into the woods. They crossed a stream, and for the first time George was scared, because he thought that the men were going to push him into the water and drown him. But soon the rush of the stream diminished. He could tell that they had entered the woods from the stillness of the air and the light leaking through his blindfold. They told him to stop, and one of the men jerked the blindfold from his head. They were standing in front of a pit dug into the forest floor and lined with rough boards. The men lowered George into the pit and handcuffed one arm and one leg to crosspieces so he couldn't escape. Then they covered the pit with a piece of tin, threw some dirt and twigs onto the covering, and left.

In retrospect, the Weyerhaeusers should have been more careful. Just three years earlier the twenty-month-old son of Charles and Anne Morrow Lindbergh had been kidnapped from the family's New Jersey home. Two months later the boy's body was found in a nearby grove of trees, despite the payment of a ransom. And just the week before George's kidnapping, his grandfather, John Weyerhaeuser, had died in his Tacoma mansion overlooking Commencement Bay, filling newspapers with stories about the son of the billionaire lumberman of St. Paul.

The day after Weyerhaeuser's death, nineteen-year-old Margaret Thulin read his obituary in the *Seattle Post-Intelligencer* to her husband, Harmon Waley. While serving six months in the Idaho State Penitentiary for vagrancy, Waley had become friends with a man named William Dainard, who was serving twenty years for bank robbery. After Dainard was granted a pardon by Idaho's governor, he, Waley, and Thulin began traveling together. They had talked about robbing a bank, but Dainard immediately latched onto the idea of a kidnapping. After reading the obituary, they rented an apartment in

Tacoma, and the two men started scouting the neighborhood around George's school. They weren't even planning to kidnap him when they did, but George's sudden appearance in front of their car was an opportunity too good to pass up.

Friday evening, after a frantic search by the family and the police for the missing boy, Phil and Helen Weyerhaeuser received a ransom note at their home on Fourth Street in north Tacoma. The kidnappers demanded $200,000—the equivalent of roughly $3 million today—in small, unmarked bills. As soon as the money was collected, the family was supposed to send the message "We are ready" in the personal column of the *Seattle Post-Intelligencer* under the name Percy Minnie. The ransom note said that further instructions would be given upon completion of the kidnappers' instructions. "We don't want to hurt anyone if we can get out of it," the note said, "so if you just follow the rules as they are lain down by us you will have the one you love back home in a weeks time if you care about them $200,000 worth. So just remember a slip on your part is a slip by us. Don't do it." The note was signed "Egoist."

George's abduction was front-page news across the country. Thousands of people gathered in front of the Weyerhaeusers' modest house near Stadium High School to gawk. Newspapers flew in star reporters from around the country who wrote breathless stories filled with fact-free speculation. The press interviewed George's classmates and analyzed the ransom note. They published pictures of George without his curly brown hair, since his barber was convinced that the kidnappers would have cut it off to disguise George's identity. They speculated that three members of a New York gang who were reported to be in the area were behind the abduction and wrote long stories about every suspicious-looking boat or car seen nearby.

Within a few days the Weyerhaeusers had gathered the $200,000 from relatives and bank accounts. The FBI, which had dispatched more than a dozen agents to Tacoma to work on the case, recorded as many of the bills' serial numbers as they could, and the ten-page list was distributed to railway depots, hotels, banks, and post offices.

After George's first night in the pit in the woods, Waley, Thulin, and Dainard were distraught over how much coverage the story was receiving in the newspapers, and they knew they could not keep driving back and forth to the pit without raising suspicions. They moved George to a second pit, where he again was left overnight. The third day they locked George in the trunk of Dainard's Ford coupe and drove the three hundred miles to Spokane, where they kept George chained to a tree in the woods while they figured out a better place to keep him. That afternoon, they rented a furnished apartment, telling the landlady that they were salesmen. Then they went to a store, obtained a large Uneeda cracker box, and smuggled George into the house inside the box. The two men made him write a letter to his family assuring them that he was still alive. After that, they locked him in a closet with Waley on guard as Dainard left to collect the ransom money.

By this point, George knew that Dainard was the dangerous one. He'd roughed George up on the way to Spokane, and the tough talk all came from Dainard. But Waley, with his sad eyes behind the black hood, seemed to feel sorry for the boy. He played songs on his ukulele outside the closet in which George was locked.

On Wednesday, May 29, George's father received a letter from the kidnappers instructing him to register at the Ambassador Hotel in Seattle at seven o'clock that evening under the name James Paul Jones. Included in the letter was George's handwritten note. Later that day, a taxicab driver delivered another note. It told Weyerhaeuser to drive alone in his car to an intersection near a brewery between Seattle and Tacoma. There, Weyerhaeuser found further instructions along the side of the road in a tin can bearing a white cloth. As directed by the note, he drove down the road looking for another white cloth but found nothing. He sat in his car for hours but nothing happened. Eventually, he gave up and drove back to Seattle with the ransom money.

The next morning, Weyerhaeuser received a telephone call from a man demanding why he had failed to follow instructions in the second note. Weyerhaeuser said that he had never found the second note. The kidnapper told him that he had one last chance. That night, Weyerhaeuser drove to a dirt road near the present location of the Seattle-Tacoma Airport. There, he followed instructions in a progression of tin cans along the side of the road. The final note told him to place the money in the front seat, leave the car running, the dome light on, and the driver's door open. Weyerhaeuser obeyed the orders and, after walking a hundred yards down the road, saw a man leap from the bushes, hop into the car, and speed away. Weyerhaeuser walked back to the highway and caught a ride to Tacoma.

At 3:30 a.m. on June 1, after eight days in captivity, the kidnappers released George onto a dirt road in the woods east of Seattle with two dirty blankets and a single dollar bill. George wandered six miles down the road until he found Louis P. Bonifas's farm. He walked to the back of the house and knocked on the door. "I'm George Weyerhaeuser," he said. Bonifas called the police department to inform them that George was safe and then began to drive him to Tacoma. On the way, John Dreher, a sports writer for the *Seattle Times,* intercepted the car. Dreher misled Bonifas into believing that he was a police officer. Bonifas handed the boy over, and on the way back to Tacoma, Weyerhaeuser's second kidnapper conducted the interview of a lifetime.

Finally, at 7:45 a.m., George arrived at the Weyerhaeuser residence. Press swarmed the house, waiting for a reaction. Finally, the Weyerhaeusers' spokesman emerged and asked that the coverage come to a halt to "reduce any bad effects on his future life."

In the days following George's return, the ransom bills began to surface in Salt Lake City stores. On June 8, Thulin was arrested at a Woolworth when she paid for a twenty-cent purchase with a $5 ransom bill. At the FBI office, another bill was found in her purse. Thulin offered conflicting stories to the investigators until eventually she gave her address. Agents staked out the place that she and Waley had rented just three days before, and soon they caught Waley with several marked bills in his pocket. Upon searching the residence, they found $3,700 of partially burned bills in the stove.

Waley and Thulin both signed confessions admitting to their parts in the kidnapping. They then led the FBI agents to $90,790 of the ransom money, which they had buried near an anthill in Emigration Canyon east of Salt Lake City. However, Dainard eluded arrest for nearly a year. Finally he was captured in San Francisco by the FBI, just a week after FBI Director J. Edgar Hoover had declared him Public Enemy Number One. As the leader in the kidnapping, Dainard was sentenced to serve sixty years in the McNeil Island Federal Penitentiary near Tacoma. Soon after, he was deemed to be insane and sent to Alcatraz.

Harmon Waley received a sentence of forty-five years in prison for kidnapping and also ended up in Alcatraz, which had been opened the year before to hold violent and incorrigible criminals. Margaret Thulin, who pleaded during the trial that her Mormon faith required that she obey her husband faithfully, was sentenced to twenty years at the Federal Detention Farm in Milan, Michigan.

During Waley's time in prison, he sent George Weyerhaeuser letters apologizing for his actions and asking for forgiveness. When he was paroled on June 3, 1963—almost thirty years after the kidnapping—he wrote Weyerhaeuser another note and asked for a job in the company. George found him a place at a Weyerhaeuser plant in Oregon. When asked why he helped someone who had caused him and his family so much pain, Weyerhaeuser said, "I went through all sorts of sensations when I was kidnapped, from fear and concern to the point where I felt sorry for him. I guess I thought he had paid his debt."

• • •

People who know George Weyerhaeuser insist that the kidnapping had no long-term effect on him. Weyerhaeuser himself later said that "a nine-year-old boy is a pretty adaptable organism." He went to Yale, as had his father and grandfather, graduating in 1948, the same year as the first President Bush. He married the beautiful daughter of a prominent lumberman, and they had six children of their own.

But George Weyerhaeuser turned out to be a different kind of leader from his father, grandfather, and great-grandfather. He might ask friends for advice, but when he made a decision, he made it on his own and stuck with it. He did not need or seek consensus before moving forward. He decided what to do and did it.

When Mount St. Helens began to spout smoke and ash in the spring of 1980, Weyerhaeuser did not change the company's plans. Company crews would finish logging the land around the mountain and get out.

If Weyerhaeuser had been a more cautious or consensus-driven man, maybe things would have turned out differently. But that was not his style. As the volcano's activity intensified, he exhibited the same resolve and self-reliance that he had so many years before.

PART 2
THE WARNINGS

The largest tree Weyerhaeuser ever cut, just to the northwest of Mount St. Helens

THE MOBILIZATION

ON WEDNESDAY, MARCH 26, 1980—SIX DAYS AFTER THE FIRST earthquake rocked Mount St. Helens—three dozen or so people squeezed into the first-floor conference room of the US Forest Service Building in Vancouver, Washington, a midsize suburb on the north bank of the Columbia River opposite Portland, Oregon. At nine a.m., Bob Tokarczyk, supervisor of the Gifford Pinchot National Forest, called the meeting to order. Law enforcement officials from all three of the counties adjoining the mountain were there. The state had sent representatives from its Department of Emergency Services, which at that point mostly handled civil defense in Washington, and from its Department of Natural Resources, which was responsible for about 10,000 acres of state-owned forest to the west of Mount St. Helens. As by far the largest landowner and employer in the area, Weyerhaeuser of course had a representative at the meeting. County commissioners, state patrolmen, fire district personnel, reservoir operators, and state geologists sat at the table. And around the edges of the room were an assortment of television, radio, and newspaper reporters, all eager to tell the story of Mount St. Helens's awakening.

A few minutes after the meeting began, Tokarczyk introduced Donal Mullineaux, who had flown to Portland the night before. Mullineaux, fifty-five, balding, with dark-rimmed glasses and a scientist's wardrobe, was a geologist with the US Geological Survey who worked out of the survey's office in the Denver area. But even though he worked in Colorado, no one knew more about the geology of Mount St. Helens than he did. For two decades, he and his colleague Dwight Crandell—whom a college geology professor once called Rocky, after which the name stuck—had been spending their summers driving and walking around Mount St. Helens, looking up at the mountain, and wondering what might happen. Like the four other volcanoes in Washington State, Mount St. Helens is a stratovolcano, meaning that it's made up of layers of ash, rock, lava, and mud—a sort of geological parfait. By digging into ridgetops, hillsides, and river valleys around the volcano, Mullineaux and Crandell had learned more about Mount St. Helens than was known about any other volcano in the Cascade Mountains. Building on earlier work, they confirmed that the volcano was relatively young, having formed about 40,000 years ago. They divided its frequent eruptions into four periods, with the latest starting about 4,000 years ago. The cone of the volcano, they showed, was less than 2,500 years old—younger than the Acropolis in Athens. That's why the mountain was so symmetrical and smooth—glaciers had not yet had time to scour out deep valleys on the mountain's sides. They discovered that Spirit Lake, just to the north of the volcano, was formed when a landslide from the mountain blocked the north fork of the Toutle River several thousand years ago. They found that the town of Toutle was built on an ancient mudflow generated by the volcano.

They also learned that, over the last four millennia, Mount St. Helens had been the most active and explosive volcano in the United States. Two years before that meeting in Vancouver, Mullineaux and Crandell had published a report entitled *Potential Hazards from Future Eruptions of Mount St. Helens Volcano, Washington: An Assessment of Expectable Kinds of Future Eruptions and Their Possible Effects on Human Life and Property*. In that report, they described the evidence they had found documenting the volcano's violent past. They had uncovered a foot-thick layer of white pumice blown from the volcano to a ridge six miles to the east in about the year 1500. Five hundred years before that, a "strong laterally directed explosion" threw what are known as lava bombs more than three miles from the volcano. In their travels over the countryside, Mullineaux and Crandell found thick layers of ash from Mount St. Helens hundreds of miles away.

Over the past millennium, the mountain had erupted about once every hundred years, Mullineaux and Crandell wrote. The last eruption was in 1857—a pioneer newspaper reported that year that the volcano "has for the last few days been emitting huge volumes of dense smoke and fire, presenting a grand and sublime spectacle," and a Canadian artist named Paul Kane, who visited the area a few years earlier, painted the volcano with a red-hot eruption spewing from its northern flank. At times over the past four thousand years the volcano had gone dormant for a few centuries, but it did not appear to be in a dormant period anymore, Mullineaux and Crandell wrote. "In the future," they concluded, "Mount St. Helens probably will erupt violently and intermittently just as it has in the recent geologic past, and these future eruptions will affect human life and health, property, agriculture, and general economic welfare over a broad area." In fact, an eruption could be expected "perhaps even before the end of this century."

At the meeting in Vancouver, Mullineaux recounted this dire history for the assembled officials. He observed that a landslide could cascade down the southwest side of the mountain into Swift Reservoir and overtop the Swift Dam, triggering a flood that could devastate downstream communities. He pointed out that mudflows could travel down river valleys faster than a man could run. Most alarming of all, he mentioned the possibility of pyroclastic flows. Being in a pyroclastic flow is sort of like being in the most powerful hurricane on earth but with most of the atmosphere replaced by red-hot pulverized rock. Such a flow can travel at speeds of hundreds of miles an hour up and over ridges, down valleys, and around obstacles. People die by suffocating on the stone dust that fills their lungs, by burns from the enveloping ash cloud, or by being torn apart by the force of the blast. In 1902 a pyroclastic flow from Mount Pelée on the Caribbean island of Martinique swept down the flank of the volcano and killed nearly 30,000 people in St. Pierre. Only two people in the city survived—one a man who was in solitary confinement in the town's jail (though he was horribly burned when ash from the pyroclastic flow entered through a small grill in the door of his cell).

The people listening to Mullineaux in the Forest Service conference room didn't know what to make of his presentation. Here he was describing catastrophic events around Mount St. Helens that had occurred in the relatively recent past. Yet it was difficult to reconcile those images with the serenity of the forests, streams, and lakes around the mountain—a serenity that people in the Pacific Northwest knew well. Though Mount St. Helens is relatively isolated from the main population centers in Washington State, the area around the mountain had long been a popular place to vacation. In 1901, a prospector named Robert Lange oversaw the construction of a rough wagon road forty-six miles from Castle Rock up the Toutle River valley to Spirit Lake. Over the years the road was widened and paved, and campgrounds, lodges, and resorts were gradually built around the lake. Beyond the lake, the Spirit Lake Highway—an incongruous name for such an unassuming road—continued three miles up the mountain's flank to Timberline, where Dave Johnston would give his impromptu press conference the day after the meeting in Vancouver. A couple of miles downstream from Spirit Lake, eighty or so private cabins occupied Lange's old homestead. On a typical summer weekend, several thousand people might be scattered around Spirit Lake, with hundreds more in the surrounding woods. Many people in Washington and Oregon still remember childhood trips to Spirit Lake, the shock of jumping into the freezing-cold water, the reflection of the mountain off the lake's rippled surface, the endless stars on dark summer nights. The scene, commemorated in countless coffee-table books and Washington State calendars, seemed timeless.

The Spirit Lake Highway was not the only road in the vicinity of Mount St. Helens. Over the years, Weyerhaeuser and the US Forest Service had built more than five thousand miles of logging roads into the woods around the volcano. While many of these roads were winding gravel truck lanes, others were major thoroughfares—two lanes, paved in places, well graded, down which hundreds or thousands of cars and pickups traveled each day as the Weyerhaeuser employees went to work. Before the reawakening of Mount St. Helens, most tourists were never aware of these roads. All they saw were sets of tracks leading into the woods off the Spirit Lake Highway. But Weyerhaeuser maps of the area were—and still are—thick with a maze of easily accessible logging roads.

In the Vancouver conference room, on the basis of a few days' worth of earthquakes, Mullineaux seemed to be saying that everything around Spirit Lake and along the river valleys leading to Mount St. Helens could be destroyed almost instantly. The people in the room of course knew that Mount St. Helens was a volcano, but given that it had not erupted in more than a hundred years, just how likely were all these scenarios? Restrictions on the area around the volcano would mean shutting down the resorts and campgrounds. People would lose their incomes. The atmosphere got tense. An official from the state Department of Natural Resources asked, "You mean to tell us that we as a nation can send a man to the moon and you can't predict if a volcano will erupt or not?" That's exactly what Mullineaux was saying. Many people in the room were incredulous. "I suppose that is not an unusual phenomenon," one Forest Service official who was at the meeting said later, "when you consider how you would probably react if a total stranger showed up on your front doorstep and announced that you had an active volcano in your backyard. That's just a little hard to swallow."

Mullineaux's credibility was also up against what turned out to be a highly unfortunate precedent. Five years earlier, Mount Baker, an even larger volcano twenty miles south of the Washington-Canada border, began to spout dark clouds of ash and steam, and the amount of heat given off by the volcano increased tenfold. Geologists said that even a small eruption could flood Baker Lake on the southeastern side of the volcano, and they pointed to landslides just a few hundred years old that had inundated the sites of campgrounds. Based on their warnings, the Forest Service closed the area around Baker Lake and opened the sluice gates on the Upper Baker Dam to lower the reservoir's water level. Tourism plummeted, leading local outfitters, restaurants, and hotels to complain to the press that the Forest Service had overreacted. As the

Concrete Herald wrote on October 10, 1975, "Maybe it's time to take a lot closer look at the bureaucratic decisions being made by some of our governmental agencies and to start reducing their powers back to where the citizens control instead of being controlled." Even as the Forest Service stood firm, the volcano refused to erupt. Not until the following spring did the Geological Survey issue a new statement that "there was now no clear evidence of a forthcoming eruption." The area was reopened and the lake refilled.

Many of the problems that plagued the response to Mount St. Helens's reawakening were apparent in that Vancouver conference room. Because of the mixed ownership created in part by the land grants, many private companies and public agencies owned land around the mountain, and each had a different set of interests and authorities. Weyerhaeuser wanted to keep the land open to logging. Cabin owners, hunters, fishermen, and hikers wanted the land to stay open to recreation. The Forest Service and local law enforcement could try to restrict access, but they had to answer to politicians, who were under pressure from their constituents. And no one had authority over all the stakeholders. The best that could be hoped was that they would agree to communicate and cooperate, which was the main outcome of that meeting in Vancouver.

At the same time, Mullineaux and the other geologists who had studied the mountain found themselves in a very uncomfortable situation. Mullineaux was a scientist who had spent much of his career studying events that had occurred hundreds and thousands of years in the past. He wrote papers that were reviewed by his colleagues, carefully revised to accommodate their comments, and published in scientific journals and reports. Now he was suddenly a public spokesman whose words were splashed across headlines the next day. "I'd give them facts, but they wanted predictions," he said. "To me, they wanted things that scientists could not do." Mullineaux and Crandell never saw themselves as responsible for establishing restricted zones around the mountain. Their job, as they saw it, was to lay out the hazards so emergency planners could decide what to do. But they found themselves in a situation where everyone looked to them for direction.

• • •

By Tuesday, April 1, six days after the meeting in Vancouver, it was clear that the situation at Mount St. Helens was going to be a lot worse than what had happened at Mount Baker. On the last day of March, the seismometers at the University of Washington had picked up a particularly ominous kind of earthquake. Most earthquakes consist of a jumble of incoherent shaking motions as waves travel both through the ground and along the earth's surface. But buried deep in the traces from the seismometer on March 31 was a kind of shaking known as harmonic tremor, where the earth shakes in a kind of rhythm. Harmonic tremor is generally associated with the movement of fluids like molten magma or volcanic gases through channels inside the earth. That had never happened at Mount Baker. But starting on March 31 and continuing for the next several days, harmonic tremor suggested that magma was on the move beneath Mount St. Helens.

On April 1, a new eruption of ash and steam reached elevations of 16,000 feet, and small steam vents opened on the northeast side of the mountain. Over the next week, earthquakes and eruptions continued, mostly steam, but some with ash.

Actually, the term "eruption" is somewhat misleading in this case. When most people think of an eruption, they think of magma emerging from inside the earth and flowing over the ground (at which point it becomes known as lava). But the March and April eruptions of Mount St. Helens did not produce lava. They were caused by molten rock and gas inside the volcano heating up the mountain's surface. When the groundwater in the volcano's surface rocks got hot enough, it flashed into steam and spouted from the newly formed crater. Some geologists don't even like to call these events eruptions, preferring the term "steam explosions." But usually they are called "phreatic eruptions," from a Greek word meaning "well" or "spring." Sometimes these phreatic eruptions carried ash into the atmosphere, which gradually darkened the flanks of Mount St. Helens. But these eruptions did not appear to contain any magmatic material from inside the volcano.

The term "ash" also can be confusing. Most people think of ash as the product of something burning, like the ashes left over from a fire. But the ash from a volcano is not the product of combustion. Rather, it's finely ground-up rock produced by the energy inside a volcano. In phreatic eruptions, the ash comes from existing rock that's been pulverized by expanding gases. In magmatic eruptions, ash is formed when gas bubbles contained in the magma rapidly expand and fragment the lava, generating a cloud of gas and hot particles. Whatever the source, ash can be one of the most destructive products of a volcanic eruption. It can kill crops, contaminate water supplies, incapacitate equipment (including the engines of jet aircraft), and bring down buildings if enough of it piles up on roofs.

Though phreatic eruptions can sometimes seem like sideshows to the main event of magmatic eruptions, they can be devastating. The 1883 eruption of Krakatoa, which killed more than 36,000 people and created the loudest sound ever recorded on earth, was partly a phreatic event. So was the 2014 eruption of Mount Ontake in Japan that killed about sixty people as they desperately sought to escape a billowing cloud of ash that erupted without warning from the mountain. The phreatic eruptions of Mount St. Helens that began on March 27, 1980, were small in comparison to what was to come, but the people who saw them that spring could not fail to be impressed.

• • •

By the beginning of April, Mullineaux's partner on the Mount St. Helens research, Rocky Crandell, also had moved from Denver to Vancouver to help with the crisis. The very afternoon that Crandell arrived, Mullineaux said that what he desperately needed was a map showing the danger zones around the volcano. Crandell found a Weyerhaeuser map of the Mount St. Helens tree farm, locked himself in an office, and the next morning emerged with a marked-up version of the map. It had three zones. One showed what might happen in the case of a small eruption, such as those that occurred repeatedly during the nineteenth century; one showed a moderate eruption; and one showed a worst-case eruption as large as any in the past 4,500 years. For such an eruption, Crandell predicted that avalanches could descend as far as Spirit Lake,

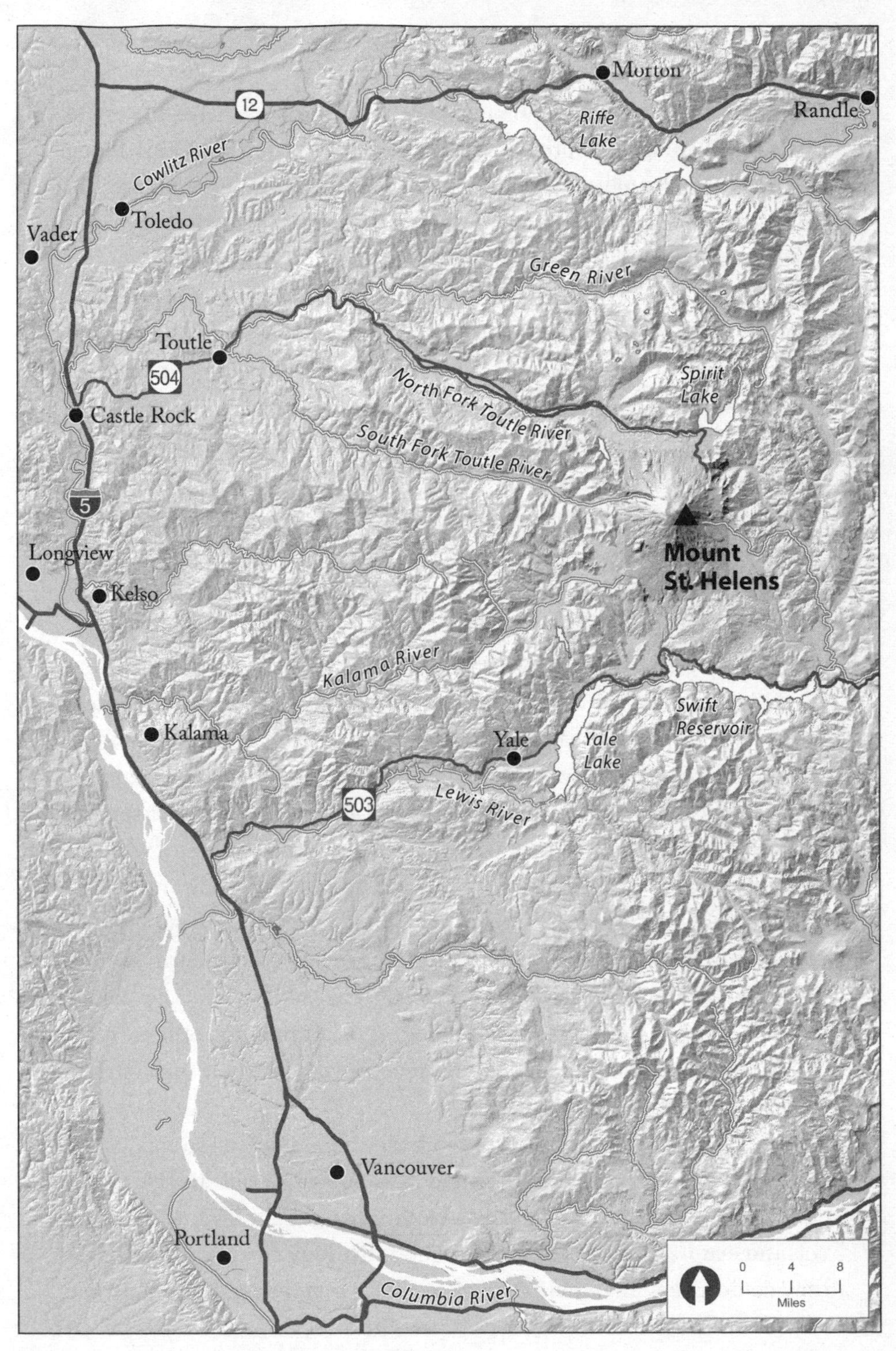

The rivers and towns west of Mount St. Helens

five miles from the summit, and that pyroclastic flows would surge at least fifteen miles down the north and south forks of the Toutle River. Ashfalls could be more than three feet deep twenty miles from the volcano and a foot deep fifty miles from the volcano. This worst-case scenario became the basis on which many future decisions were made.

Crandell's hazard maps immediately pointed to a major problem. The volcano was fast becoming a tourist attraction. People were traveling to Mount St. Helens not just from Washington State but from all over the world. Journalists came to the volcano to interview the locals and film the occasional eruptions. Television stations parked their microwave trucks—a recent innovation that allowed them to get on air almost immediately with breaking news—along the Spirit Lake Highway and in nearby towns. Local officials played up the volcano as a destination. Entrepreneurs near the mountain began to sell T-shirts, postcards, coffee cups, picture books, ashtrays made of volcanic ash, and anything else they thought visitors might buy.

Most of those who came to see the volcano went away disappointed. In the late winter and early spring, southwestern Washington State is one of the cloudiest places in the nation. The area around Mount St. Helens gets about ten feet of rain per year, and in the rainy season the mountain can be socked in for days on end. But every once in a while, if the sightseers were lucky, the clouds would part, the volcano would emerge, and a puff of steam-driven ash would rise from the peak. Amateur photographers began to set up on ridges and in clearings around the mountain where they could take photographs when the clouds lifted. Some began to sell their photographs, which got others interested. Now people had a commercial reason, besides their own curiosity, to come to the mountain.

While some of the locals profited from their souvenir stands, many were annoyed by the mountain's sudden popularity. State and local officials were telling them that they would have to get out quickly if the volcano erupted. Yet as the roads filled with tourists' cars, they knew they couldn't get anywhere fast. "If we had to move a lot of the people out of here, how would we get past the sightseers?" asked one resident of the town of Yale. Tacoma's *News Tribune* quoted an emergency official saying, "People are swarming in from all over, putting their lives in danger. . . . Sunday, when the weather was clear, the road up to the mountain looked like downtown Seattle at rush hour." The airspace over the mountain was almost as bad. One afternoon, seventy aircraft full of media and sightseers were circling the mountain.

• • •

On March 27, when the first ash clouds rose from the mountain, law enforcement officials from Cowlitz County, in which the western approaches to Mount St. Helens are located, had set up a roadblock on the Spirit Lake Highway. But this was an odd kind of roadblock. It was designed to keep some people out but not others. For one thing, Weyerhaeuser was still logging the woods between the roadblock and the mountain, so Weyerhaeuser employees could go through. Scientists studying the mountain were granted free passage, as were journalists if they had permission to do a story closer to the mountain. Even homeowners, at the beginning of the crisis, could talk their way through if they made a convincing case to whoever was manning the gates.

Furthermore, the roadblock wouldn't stay put. As warnings intensified, it would move farther away from the mountain. But then property and business owners would put pressure on public officials to move the roadblock back up the valley. Four times it moved up or down the Spirit Lake Highway, finally coming to rest about twelve miles down the road from Spirit Lake.

Even more important, the roadblock was only on the paved road. None of the logging or Forest Service roads that threaded the woods were cordoned off. In fact, a major gravel road paralleled part of the Spirit Lake Highway, since fully loaded logging trucks would quickly have demolished the highway's asphalt. Many of the local people had used these gravel roads for years—to get to work, to go fishing or hunting, or just to get away for a day or two. Most households had pickups to drive those roads—as they do today—Ford F-250s and Broncos, Chevy Silverados, Dodges, even the occasional Toyota or rotary-engine Mazda. And many people in Washington and Oregon are used to driving on gravel roads to go camping, hunting, fishing, or hiking. Soon maps surfaced showing tourists how to take the back roads to get close to the mountain. Eventually the roadblocks became so expensive to man and so easily evaded that law enforcement officials didn't even bother to station officers there. Anyone who wanted to get past the roadblock had no trouble doing so. Mount St. Helens was becoming even more popular as an incipient disaster than as a vacation destination.

IN THE VADER TAVERN

EVERY DAY AFTER WORK, JOHN KILLIAN DROVE PAST THE VADER Tavern on the way to his home a few blocks from the Seattle-Portland rail line. Sometimes he stopped in for a beer, though not so much after he and Christy got married—he enjoyed his evenings at home with Christy too much to live like a bachelor again. But every once in a while he would stop in for a drink with his friends and coworkers. He needed to catch up with the news about the volcano.

In the spring of 1980, John was working for Weyerhaeuser on a road-building crew. Twenty-nine years old, lean and bushy-haired, he had lived in Vader almost his whole life, except for his time in the navy. His parents' house was just a block from his own, and all three of his sisters lived in the area, along with his cousins and aunts and uncles and in-laws and other relatives. The Killians had been living in and around Vader for more than half a century, since John's father had moved there as a boy. When John and Christy had children, they would raise them there too.

Before it burned to the ground in 1988, the Vader Tavern sat on A Street between two vacant lots where previous buildings had burned. The tavern was almost one hundred years old, with a magnificent solid-oak stand-up bar that had come around Cape Horn in the 1880s. The building had a false front, as did many of the remaining commercial buildings in Vader, and a sloped roof over a low wooden porch, so that it seemed more like a set for a western movie than a small-town tavern. People came from miles around to look at their reflection in the back bar, to watch the cascading neon beer signs, and to drink Rainier on tap, or Schlitz out of cans, or Coors all the way from Colorado if they were feeling rich.

Every small logging town around Mount St. Helens had taverns where the loggers met to drink and talk after work, and this is where, in April and May of 1980, the loggers argued whether it was safe to be working so near the mountain. Weyerhaeuser had moved many of its crews onto the hillsides and ridgelines so they wouldn't be stuck in a valley if an eruption caused a mudflow. But some of the crews were working less than five miles from the mountain, and if anything the logging seemed to have intensified that spring, despite the economic downturn. Ash from the volcano mingled with the April snows, yielding a slushy gray mess. When an especially strong earthquake struck, the loggers would have to lower their chainsaws and lean against a tree trunk to remain standing. As usual, April was a cloudy month around the mountain. But every once in a while the clouds would momentarily part and the loggers would gaze at what the mountain had become. Like Mounts Rainier, Baker, and Hood, Mount St. Helens was named by the English explorer George Vancouver for someone he wanted to honor—in this case, an English diplomat named Alleyne FitzHerbert who was given the title of Baron St. Helens after forging an important treaty with Spain. FitzHerbert never saw the mountain, which perhaps is just as well, since his tenuous connection with the mountain and unusual title further reinforced its distinctly feminine aspect. The Indians in the area, whose ancestors had watched Mount St. Helens periodically erupt for many thousands of years and generally stayed away from it, always gave it female names: Loowit, which means "keeper of the fire"; Lawetlat'la, which means "the smoker"; or simply Si Yett, which means "the woman" in the Penutian language of the Yakima. Their legends spoke of an old woman whom the gods made young, reflecting the mountain's geologic past. But now the mountain's feminine nature was disfigured, as if an evil gray liquid had leaked from its tip, and on the north side of the mountain a strange disfiguration had appeared, a malignancy on the once-perfect cone.

The loggers at the Vader Tavern were of two minds about the mountain. Some were obviously worried about it. They didn't say so very loudly, as they raised their beers to drink, for fear of being labeled cowards or shirkers. But they'd read enough in the papers and seen enough on television to know that if a major eruption occurred, they could be in serious trouble. Many of the loggers were familiar with the mountain's past. Whenever a Caterpillar cut into a hillside to lay down a new roadbed, the roadcut would reveal multiple layers of ash. Some were thin, but others were a foot or two thick, with no seams to indicate that they were the product of more than one eruption. If the loggers were caught in an eruption like that, it would be difficult to escape. Their supervisors said they would have time to get away. But they'd seen the ash clouds billowing from the top of the mountain, and they'd heard the crackling and groaning of the mountain's explosions. They wouldn't be able to get away if they were caught inside one of those.

John and Christy had talked about it. She drove a forklift at Weyerhaeuser's Green Mountain mill outside Toutle and was far enough away from the volcano not to be worried. But John was working just a few miles northwest of the peak. If the mountain blew, she wouldn't know what had happened to him. John shrugged off her concerns. He'd been a boilerman in the navy, serving on the USS *Badger*, mostly off the coast of Vietnam, where the ship and its crew were assigned to cover operations on land and keep an eye on the Russian submarines patrolling in the South China sea. He knew every one of the 416 valves in the boiler room by heart. He was good around machines after a lifetime in the woods. They'd been fired upon lots of times in Vietnam, but he'd never come close to getting hurt. If the mountain did anything, he and his buddies would figure out how to get away.

Besides, they both knew that logging was a dangerous business. He and his coworkers had to take occasional risks in the woods; they had no choice, even though the cemeteries were filled with men who paid for those risks with their lives. The year after he graduated from high school, Herschel Bowers was killed when a limb fell out of a tree and ran him through the neck. Tom Galt was killed the next year when a crane hit a power line while he was holding on to a choker cable. Jim Mason, Richard Caywood, and George Rockett were all killed in 1975 by falling trees—that was a bad year. Ben Snow got run over by a bulldozer the next year, George Hack got crushed by a rootwad when he was bucking a windfall, Dee Bond was killed just last year felling timber right by Vader.

And those were just the men who'd been killed. Almost everyone working in the woods had been injured at one time or another. It was like the army guys in Vietnam—they just hoped when it happened that it would be enough to keep them out of the action but not too bad. John knew that Christy was aware of the risks—everyone in Vader was. But they couldn't spend their lives worrying about what might happen.

• • •

In 1910, Vader, Washington—then known as Little Falls—was the largest city between Portland and Tacoma, with almost 5,000 residents. It sat in a broad valley surrounded by heavily forested hills that stirred the blood of every lumberman who saw them. From openings in the forest emerged gigantic trees, shorn of their limbs and bucked into railcar-sized lengths, which were loaded one per car to travel on the Northern Pacific line to mills in Tacoma or Portland. Most of the money generated from those trees went elsewhere, but enough stayed in Vader to create a brief period of prosperity.

The town had four hotels—the Stillwater, Spangler, Shelhorn, and Bannon—an opera house that doubled as a boxing arena, a hospital, five saloons, and a surrounding phalanx of churches. The largest building in town was the Little Falls Fire Clay Company, where 100 men made clay chimney pipes, flue linings, paving bricks, drain tile, and sewer pipe. The Stillwater Logging Company employed another 180 men at its sawmill by the rail line and at three logging camps in the woods. People looking for good jobs in factories or logging in the woods flooded into the town, including many from the depressed logging states of North and South Carolina.

In 1913 the Northern Pacific objected to the name of the town, saying that a Little Falls already existed on the company's rail lines in Minnesota. For a replacement, the townspeople chose the name of a popular resident, Martin Vader. But the new name seemed to bring ill fortune. That same year, the Weyerhaeuser Company bought the Stillwater Logging Company and shut down the sawmill, laying off hundreds of workers. The next year the brick factory burned to the ground, leaving only its smoldering chimneys and kilns.

Soon the town entered into a dark period. The townspeople began buying insurance for their homes and then setting them afire. A resident described in a letter approaching the town one night and seeing the distant glow of burning houses. The hotels and saloons closed, and then the shuttered churches began to burn.

In 1929, John Killian's father, Ralph Killian, moved to the vicinity of Vader from North Carolina as a nine-year-old boy. The town was already in decline, but the Killian family endured, partly because they were surrounded by other families that were facing the same struggles. During World War II, Ralph was a first-class petty officer in the navy, after which he returned to Vader and took a job with Weyerhaeuser. It was a good job, a union job, and he moved slowly up the ranks, from choker setter to rigging slinger to crew boss. He was a tough boss but fair, with a streak of sympathy for the men who were doing the jobs that he used to do.

Meanwhile, the town of Vader continued to fade. Over time it took on a curious gap-toothed appearance, with just a few houses still standing on each large block. It still had a few businesses—a general store, the tavern—but the only place left to eat was the Little Crane Café, which occupied the sole remaining structure left from the Stillwater Lumber Company, a square shack right next to the railroad tracks that used to house tools and spare parts for the mill. Around the town the hills acquired the curious shorn look of many logged areas in the West. Traditionally, such hillsides are said to look like they've had a bad haircut, but actually it's more like a haircut by a madman: the boundaries of cutover regions have no obvious design or rationale, with only the occasional single standing tree to remind onlookers of what once was there.

As businesses and houses continued to burn, the town's population dropped—from 5,000 to less than 1,000 by 1980. Nevertheless, the Killians stayed on. Other members of Ralph's family from North Carolina joined him in Vader, and they married into local families and began to have children. Over the decades an extended network of interrelated families took shape not just in Vader but in the nearby towns too. No matter where you went in Vader, Toledo, Ryderwood, Winlock, Castle Rock, or Toutle, everyone knew the Killians and their kin. They were part of the country around there just as much as the hills and streams were.

THE BULGE

IN APRIL THE GEOLOGISTS MONITORING MOUNT ST. HELENS, LIKE the loggers working around the mountain, realized that the shape of the volcano was changing. A large bulge had formed on the north side of the mountain. It was about a mile from top to bottom and a little over half a mile across. It stretched and fractured the glaciers on the side of the mountain, so that they took on a curious blistered appearance, and it continued to grow. Each day, the bulge expanded outward by about five feet. "A major deformation like this is pretty extreme," Dave Johnston told a newspaper reporter. "It surprises us, to be quite honest. It's not that it is totally unexpected, but it is a pretty major change in the shape and size of the volcano."

By this time, an eclectic group of geologists had gathered in Vancouver and other towns around the volcano. Several volcanologists who had worked at the Hawaiian Volcano Observatory on the Big Island flew to Washington State to help monitor Mount St. Helens, including Don Swanson, Jim Moore, and Pete Lipman. Bob Christiansen from the Geological Survey's Menlo Park, California, office oversaw the monitoring effort, while Dan Miller and Rick Hoblitt from the Denver office assisted Crandell and Mullineaux. Richard Waitt analyzed ashfalls, and Thomas Casadevall helped Johnston track volcanic gases. Meanwhile, other geologists and seismologists not associated with the USGS came to southwestern Washington to inspect the volcano and make predictions about what could happen.

Since his press conference the day after the first eruption, Johnston had been more worried than many of his colleagues about the dangers posed by the volcano. He and a few of the other geologists had begun talking about the similarities between Mount St. Helens and another volcano, Mount Bezymianny, which is in Russia on the Kamchatka Peninsula a thousand miles northeast of Japan. In September 1955, earthquakes began to shake Mount Bezymianny, and three weeks later the mountain erupted just as Mount St. Helens had erupted at the end of March, with steam and ash as the volcano heated up. Then the mountain quieted for several months, though the earthquakes continued. Suddenly, on March 30, 1956, the volcano violently exploded, devastating areas almost twenty miles away. But what was unusual about Bezymianny is that the volcano did not explode straight upward, as in the conventional image of an erupting volcano. It exploded sideways, ripping the side off the mountain and destroying everything in that direction while leaving the rest of the countryside unscathed. If something like that happened at Mount St. Helens, Johnston told the other geologists, the destruction would be immense.

The geologists had been monitoring the volcano from the Timberline parking lot on the north flank of the mountain, but they needed another vantage point if Timberline became unsafe. Early in April, a state snowplow cleared a path along logging roads to an old landing above Coldwater Creek, where the geologists erected a canvas tent on a wooden platform. As early spring snows continued to fall on the ridge, they warmed themselves around a fire barrel that Ralph Killian had donated. They called it the Coldwater observation post.

By the second week in April, Rocky Crandell was sufficiently worried about the north slope of the volcano that he decided he needed the help of an expert. He talked with Barry Voight, a respected landslide geologist at Penn State University, and asked if he would come take a look. Voight was a member of an unusual family. One of his brothers was the actor Jon Voight, whose daughter, Angelina Jolie Voight, was four when Mount St. Helens began acting up. His other brother, who went by the stage name Chip Taylor, was a musician best known for writing "Angel of the Morning" and "Wild Thing." But Barry Voight was a scientist through and through: meticulous, thorough, and sometimes brusque in presenting his results and opinions.

On April 10, Voight flew to Portland, rented a car, and drove to Vancouver, where he attended the regular morning meeting of the geologists to discuss the volcano's behavior. The next day he drove up the Spirit Lake Highway and set up a camp about a half mile east of the Coldwater observation post. He sat down on a bedroll with a notepad and binoculars and began to study the mountain. It was a fearsome sight. Gigantic cracks pierced the ash-covered glaciers on the north and west slopes. Something was causing the mountaintop to split, but whether it was magma pushing up from below or deformation of the rocks caused by the heated groundwater, Voight couldn't tell. On his notepad, he sketched out some hypotheses of what might happen. If the bulge gave way, could a large avalanche occur? He did a quick calculation of the size of a potential landslide. It would be huge, so large that it would easily reach Spirit Lake. He wrote a note to himself: "Recommend not opening the Spirit Lake area." And then "Surge wave in lake, if rockfall reaches lake, runup on far shore."

What if pressure was building up inside the mountain and a landslide suddenly released it? Voight wondered. Could a pyroclastic flow follow in the same direction as the avalanche? Such an eruption could be similar to that of the Mount Bandai volcano in Japan, which exploded sideways in 1888, killing more than 450 people and devastating the region to the northeast of the volcano.

Voight stayed on the mountain for two days, finally breaking camp when the clouds and rain returned. The next day he drove back up to the Timberline parking lot, where he told the geologists working there that they were in a very dangerous spot. If a landslide came down the north side of the mountain, it could kill them all and unleash a pyroclastic flow directed right at Spirit Lake. Voight spent a few more days in Vancouver and around the mountain, helping with various tasks and telling anyone who would listen about the landslide dangers. He posted on the walls of the office a scaled drawing of the mountain's cross-section, showing the huge prospective landslide and its relation to the observed surface cracks. Then he returned to Pennsylvania to write a report on what he had observed. "The principal hazard at Mt. St. Helen's during the period of my observations (11–19 April, 1980) involved the potential instability of the north slope," he wrote. If the slope gave way, the resulting landslide "could be as much as a kilometer thick and involve a cubic kilometer or more of rock and fragmented material." Moreover, the release of the underlying pressure from the avalanche "would likely promote further explosive activity (flashing) in hydro-thermal systems occupying the core of the volcano, the surrounding porous edifice, and perhaps also in shallow magma chambers. A catastrophic event of the kind observed at Bandai-San . . . must be regarded as a legitimate possibility."

Voight's observations and recommendations generated intense discussions among the geologists. But in the end his speculations about a lateral blast were treated as just one more possible scenario. Everyone had ideas about what could happen at the volcano. They argued their positions at the daily scientific meetings in Vancouver, but without more data, no viewpoint could prevail. The historical record compiled by Crandell and Mullineaux was still the primary source of information for the geologists. But their careful research could be misleading too. They had reconstructed the volcano's history over several thousand years, but Mount St. Helens was a very young volcano, and during its short life it had exhibited many kinds of behavior—sometimes it oozed hot lava down its flanks, sometimes it sputtered and smoked for decades on end, and sometimes it erupted explosively. Mullineaux and Crandell based their forecasts on what their studies had revealed about the past. For Mount St. Helens, however, the past was not necessarily a reliable guide to the immediate future. As Crandell told a newspaper reporter in April, "Mount St. Helens has done so many things in the past that hardly anything would be a surprise. The only thing it hasn't done is blow itself apart."

The same day that Voight set up his tent and folding chair near the Coldwater observation post, the director of the Geological Survey, a man named Bill Menard, came from Washington, DC, to see the mountain for himself. After a briefing in Vancouver, Mullineaux and a retinue of geologists proceeded to the Timberline parking lot. For the last few weeks, with the cloudy weather and all the work he had to do in Vancouver, Mullineaux had seen very little of the volcano. But standing with Menard on the north flank of the volcano and looking up to the bulge, he was shocked. He knew the volcano better than anyone in the Timberline parking lot. The geological forces that it must have taken to produce that bulge were immense.

No one should be here, he thought. "I had been there many times, but from the [aerial] pictures I had not understood how much change had taken place and how threatening that thing looked. It looked like a failure was about to occur. The bulge was terrifying. I wanted to turn around and leave."

• • •

By the end of April, the bulge had become the geologists' greatest concern. An April 25 press release from the Forest Service revealed that the bulge had raised the side of the mountain 300 feet above its original contour and was continuing to expand. It could not go on growing forever. Eventually it would have to collapse down the north side of the volcano. Crandell calculated what the failure of the bulge could do. It would flow over Timberline and into Spirit Lake, where it would generate a large wave that would slosh up the far side of the basin and drown anyone still around the lake. The landslide and water displaced from Spirit Lake then would course down the north fork of the Toutle River, sweeping away whatever it encountered. "The probability of an avalanche cannot be quantitatively determined, but such an event must be regarded as a distinct possibility and probably the most serious hazard yet posed by the current eruption up to this point," Crandell wrote in a letter to forest supervisor Bob Tokarczyk.

Other geologists were more guarded in their warnings. A large avalanche from the bulge "would be a pretty extreme scenario," Dave Johnston told the Associated Press in an article published on April 29, "but it could happen. It could be a catastrophic release of a very large volume, in which case I wouldn't want to own a house up there, or it might be a piecemeal bit-by-bit collapse that would be dangerous primarily on the flanks of the volcano." Yet Johnston and his colleagues were still working in the Timberline parking lot, which certainly would be wiped out by an avalanche of any appreciable size. Johnston was even flying into the crater occasionally to collect gas and water samples, after which he would sprint back to the helicopter and quickly fly away.

Besides recognizing the dangers posed by the bulge, the geologists in Vancouver were aware that an avalanche could trigger an even greater disaster. On May 6, for example, the Tacoma *News Tribune* carried an article featuring an interview with a local geologist named Jack Hyde, who was not associated with the Vancouver group. "I have a gut feeling . . . that as the bulge continues to grow, something dramatic is going to happen," Hyde said in the article. If the north slope gave way, a "spectacular" explosion could occur without warning. "The blast from the explosion of a Soviet volcano in 1956 blew down trees 15 miles away," he said, adding that Mount St. Helens could do the same. When Hyde was asked by the interviewer about the geologists watching the mountain from nearby observation posts, he said, "I hope they're not in a direct line. That's like looking down the barrel of a loaded gun."

Yet the geologists in Vancouver continued to monitor the volcano from its northern flank. Partly they were used to taking risks. It was their job to warn others if an eruption began. The closer they were to the volcano, the faster they could issue an alarm.

The geologists also had different experiences with volcanoes. Many had worked previously at the Hawaii Volcano Observatory, which is perched on the edge of the Big Island's Kilauea volcano. The volcanoes in Hawaii typically erupt a very different kind of lava than do the Cascade volcanoes. The basaltic lava in Hawaii, which is derived from a hot spot beneath the Pacific tectonic plate, has the consistency of raw honey when it erupts. Fed from summit vents or rifts that split the flanks of volcanoes, the lava flows downslope in pulsating streams. Explosive eruptions are infrequent, and the risk to human life from such eruptions is low.

The magmas that have created the Cascade volcanoes generally have a different consistency. As they make their way to the surface of the earth from the subducting oceanic plate, they often lose iron- and magnesium-rich crystals and gain silica from the melting of continental rocks. By the time the magma erupts, it has a consistency more like road tar than honey. This kind of magma can form stiff, sticky plugs in volcanic vents that bottle up the gases and pressure in a volcano. When the blockage gives way, explosive eruptions can result. The Hawaiian volcanologists had studied other kinds of volcanoes, but they were much more familiar with gradual eruptions of red-hot basalt than they were with explosive subduction-type volcanoes.

However, the volcanologists in Vancouver did share one belief, which gradually made its way into the public consciousness. They thought that some kind of warning would precede a major landslide or eruption. The bulge would start slipping, or earthquake activity would pick up, or sulfur emissions would increase, or new hot spots would appear on the volcano's flanks. As Mullineaux said, "There is a good chance we would have warning before a serious eruption" (though he added that "a slide could go without any warning signals"). By the beginning of May, the volcano was being monitored in several different ways, any of which might give warning of an impending disaster. Geologists were measuring the movement of the bulge from the Timberline parking lot. Seismometers around the mountain were feeding their signals to Steve Malone and the other seismologists at the University of Washington, who were conveying the information to Vancouver. Sensitive gravity stations and magnetometers were seeking subtle changes inside the earth. Planes were flying overhead and taking photos when the weather was clear, and some of them were equipped with infrared cameras to take images of hot spots on the mountain. Dave Johnston and other geochemists were looking for sulfur in the gases and water being released from the mountain. The warning that scientists expected to get from these measurements might not be much—maybe just a few hours or minutes. But it would give the people around the mountain at least a head start on some of the more dire scenarios discussed during the morning staff meetings.

• • •

On the wall of his office in Kelso, Cowlitz County Sheriff Les Nelson had a drawing of a broken-down cowboy with a caption reading "There's a hell of a lot of things they didn't tell me about this outfit when I hired on here." Nelson looked at that drawing often in the spring of 1980. Certainly if he ever ran for sheriff again, he'd make sure it wasn't in a county next to an active volcano.

Nelson had the job of keeping the people in his county safe; he spent most of his time dealing with people who'd found ways to be unsafe. He'd been a ranch foreman in Colorado before moving to Washington, where he went to work in the logging camps around Mount St. Helens. He'd gotten into sheriffing because he liked people and people liked him, and he thought of himself as a lot more sensible than many of the people in his county, so it made more sense to have him watching over them rather than the other way around. But compared with the garden-variety car crashes, drownings, and shootings he dealt with every day, this volcano was in a completely different category. He'd sat in meetings with the geologists and heard them describe the avalanches and mudflows coming down the valleys and the clouds of hot asphyxiating gas. The other people in the meetings—the people who worked in offices all day—didn't seem to take the warnings seriously, but Nelson had seen what two tons of metal can do to a person, much less a wall of mud twenty feet high.

Everybody was unhappy with him and his officers. At the barricade on the road to Spirit Lake, drivers would yell at them, "This is a free country." "I have a right to go up to my property." "I'm going to call my attorney." The restaurant and motel owners on the far side of the barricade howled about government taking away their freedom while the owners on the near side raked in more money than they had ever seen in their lives. The state hadn't made things any easier. It had closed the Spirit Lake Highway, but it hadn't closed any of the surrounding logging roads. All people had to do was buy a map from a gas station or convenience store and they could go anywhere they wanted. The sightseers weren't breaking any laws, no matter how foolish they might be. There was nothing Nelson or his deputies could do about it.

The geologists weren't much help. He'd once told a journalist, "Trying to pin down a geologist [is] like trying to corner a rat in a rain barrel," and even though that was probably too harsh, he thought it at the time and he still thought it. They were being way too cautious. If the geologists would just say that these people were going to die if that bulge gave way, at least they would be telling the truth, and he wouldn't have to be the one to say they were putting their lives in danger. Of course, people wanted to put their lives in danger—every sheriff knew that. But when the inevitable happened, and someone got hurt, there'd be hell to pay.

Besides, the press was worse than the geologists. They were the ones flying helicopters onto the crater rim like they were descending on some accident scene to interview the survivors. Sometimes they helped by scaring people with stories about red-hot lava and ash clouds that would bury a person in his tracks. But as often as not their stories just made the more adventuresome types want to get closer to the volcano. And word was getting around that money could be made by taking photographs of the volcano from as near as a person could get. Nelson had seen cars with Seattle and Bellevue license-plate frames that were filled with camera equipment turning off the highway onto logging roads, and he knew they could drive all the way to the flanks of the volcano without interception. It was making the logging truck drivers crazy, because they'd be barreling downhill on a narrow, crumbly dirt road and would suddenly encounter a station wagon crawling in the opposite direction straight from some suburb. Somebody was going to get killed out there, the truck drivers said, and it wouldn't be their fault.

But the worst thing about the press was their love affair with the outlaws. Reporters were making heroes out of the people who were loudly defying Nelson and his deputies, which was making it impossible for Nelson to do his job. An octogenarian named Harry Truman was by far Nelson's biggest problem. From the day after the first eruption, the press had been writing stories about Harry refusing to leave his lodge on Spirit Lake. "I stuck it out 54 years, and I can stick it out another 54," Harry had loudly proclaimed, a glass of Coke and Schenley's whiskey in his hand. He might have left if the papers hadn't written stories about him. But now he had his image to uphold.

Nelson felt some sympathy for Truman. He'd been running the Mount St. Helens Lodge since he gave up bootlegging in the 1920s, not long after his father was killed in a logging accident near Chehalis. He'd built that lodge up from almost nothing. He and his wife, Eddie, hosted hundreds of people each summer. He'd become friends over the years with famous people, including Chief Justice William O. Douglas and the Burlington Northern officials who stayed at the lodge each summer, since the lodge was built on land leased from the railroad.

But five years earlier Eddie had suddenly died in the lodge after going upstairs to take a nap, and that had broken Truman. He was so angry and mean that people around the lake tried to avoid him. The lodge stank of piss from his sixteen cats. The papers couldn't print half of what he said, since mostly what he did was swear. Still, journalists loved to take pictures of him drinking in his kitchen, or sitting at his player piano, or gesturing at the mountain above his lodge. They'd fly in to the lodge to interview him and bring him more Schenley's, and some reporters had told him that they'd bring him out in a helicopter if the mountain blew. Hell, look at that mountain, Nelson thought. If that thing gave way, Truman wouldn't have time to get one pant leg on before he'd be dead. There couldn't be a worse place to be. Truman thought the trees were somehow going to block all that rock and snow from descending on him, as if he'd never seen an avalanche up there. He'd told the press, "The mountain will never hurt me. When you live someplace for 50 years, you either know your country or you're stupid."

The problem was, if Nelson let Harry Truman stay at his lodge, then everyone else would point at Truman and say, Well, why not me too? The press and the people making money from selling their photos would just tell Nelson that they were doing their jobs. There was no law against that. Already he and the other sheriffs in the area were getting huge pressure from the cabin and resort owners on Spirit Lake who wanted to get to their properties. They couldn't make an exception for Truman and not make an exception for other people.

But it wasn't just Truman, Nelson knew. It was the loggers too. Hundreds of loggers were up there every day working, though many of them weren't happy about it. With the loggers in the woods, it didn't make sense that other people couldn't go in. No real laws covered the volcano. People were making it up as they went along.

It couldn't last, Nelson knew. Something had to give. As he told a reporter at the end of April, "I never wanted to be wrong about anything more in my life, but there's pressure being built up inside that mountain, and it's got to come out somewhere."

WORKING IN THE WOODS

IN THE SPRING OF 1980, JOHN KILLIAN'S JOB WAS TO SET CHOKERS behind a Caterpillar on a road-building crew about ten miles northwest of Mount St. Helens. Setting chokers is one of the toughest and most dangerous jobs in logging. Your job is somehow to get a metal cable as thick as a man's thumb around the trunk of a fallen and de-limbed tree. Once the cable is looped around the tree, you attach the other end of the cable to a separate line that leads to a clearing in the forest. You signal with a whistle or walkie-talkie to the rigging slinger, and the tree is dragged by a winch or a Caterpillar to the landing to be loaded onto a truck. Then you do it all over again.

Ken Kesey has a great description of choker setting in his novel *Sometimes a Great Notion*. Lee, the bookish, high-strung younger brother, has returned to his childhood home on the edges of the great coastal forests of Oregon to help with a massive fall harvest:

> The next log has fallen on a clear, almost perfectly level piece of ground. Unhampered by vines or brush, Lee reaches the log easily.... But the very flatness of the ground beneath Lee's log presents a problem; how do you get the cable under it? Lee hurries along the length of the big stick of wood all the way to the stump,

> then crosses and hurries puffing back, bent at the waist as he tries to peer through the tangle of limbs lining its length. . . . But there is no hole to be found: The tree has fallen evenly, sinking a few inches into the stony earth from its butt to its peak. Lee chooses a likely place and falls to his knees and begins pawing at the ground beneath the bark, like a dog after a gopher. . . .

By lunchtime, Lee is thoroughly exhausted:

> Lee allowed himself to sink to the ground beside a stump. . . . [He] had removed the heavy shirt to let the breeze dry off his sweat; the sweat wasn't much affected as he tugged, jerked, and hauled the unwieldy cable through a miasma of berry vine and fire slashing, but within a half an hour both arms were quilted from glove top to shoulder with a pattern of welts and scratches. The view he had of his stomach made him think of fabric instead of flesh, a bright garment of patchwork skin stitched together with thorns. . . .
>
> He felt sick. . . . He wouldn't move from the spot, though his leg was twisted painfully beneath him, though those bastard carpenter ants, big and shiny as carpet tacks, crawled through his shirt and across his sweating belly, and though he was sitting in a thicket of what was surely poison oak—what else?

John Killian worked hard for the Weyerhaeuser Company. One evening, after setting chokers all day, he told his mom, "I did good for George today." Except for his time in the navy, John had worked most of his adult life for Weyerhaeuser. It was the logical job for him to take. His father was a twenty-year veteran of the company. Jobs were plentiful back then, when the big trees up by Mount St. Helens were just waiting for the loggers and their saws. All he had to do was go down to the union hall and sign a few papers and he was as good as hired. Pretty much everybody around Vader went to work in the woods or in the mills, whether for Weyerhaeuser or for another company.

But by the spring of 1980, John could tell there was no future working for Weyerhaeuser. The old growth was running out fast. They'd have it all sawed down within a few years. Then the number of jobs in the woods would start falling. The old-timers like his dad would hold on to the best jobs, which meant there'd never be a chance for John to move up to rigging slinger or yarder operator. He'd be stuck setting chokers for years, assuming he could even hold a job, and not many people lasted in this job for long. He'd get hurt, or his body would wear out, or he'd just get tired of clawing through the snowberry and sword ferns and blackberry brambles.

John was making good money working a union job for Weyerhaeuser—$22,000 a year, if he could work the whole year—and Christy was making decent money driving a forklift in the Green Mountain mill. (Adjusted for inflation, John would be making more than $50,000 today.) But that wouldn't last. They were trying to have a baby. Christy would probably be pregnant soon and would have to stop working. And he could tell that Weyerhaeuser was going to start cutting back on union employees and transfer more of the work to independent contractors. It was cheaper for the company to hire gyppos and not pay union wages and benefits. Already, most of the fallers working on the ridges around the mountain were contractors. It was the way the business was going.

John and a friend had talked about going out on their own as contractors for Weyerhaeuser. You could make more money, if you worked hard and if there was work to be had. And he would need the money to do everything he wanted to do. He'd saved enough to buy some land out in the woods by Toledo. But he'd need more to start building a house, even if he used his friends to get the foundation poured and the framing done.

It was a risk, leaving the security of a union job behind. But John was used to taking risks. Once he and his friend Al Bates had been hunting elk west of Vader, up in the second-growth forest by the Grays River divide. He'd followed a buck and two does upriver, keeping the stream between him and the elk so they wouldn't hear him. The land headed steadily uphill, John on one side of the stream and the elk on the other, and the streambed narrowed and dropped into a deep gorge. It got rocky, and John had to scramble to keep the elk in sight and his rifle out of the mud. Then the elk veered away from the stream, and John knew that he'd lose them unless he got across. But the stream was now at the bottom of a deep canyon so narrow that the walls on either side almost touched. He'd have to jump it. He looked for a place where the canyon walls were close together and quickly found a spot. It should be easy to get across, he figured, just eight or nine feet of empty air. He threw his gun to the other side—now he had no choice. He backed up to get a running start, hurrying, since he knew that the elk were getting farther away. He sprinted toward the canyon, worrying just for a moment when his jumping foot landed two feet short of the edge, because that meant he'd have even farther to jump, and then he was up in the air, with nothing below him for at least a hundred feet. It was an odd sensation, as if he were flying, as if he were weightless while the world turned beneath him. But just as quickly he landed on the other side, dropping to one knee. Breathing hard, he picked up his gun and hurried after the elk.

THE RED AND BLUE ZONES

BY THE MIDDLE OF APRIL, THE FOREST SERVICE REALIZED THAT IT needed a better way of controlling access to areas around the mountain. The area above Timberline had been closed since the first earthquakes, and a roadblock on the Spirit Lake Highway kept people off the one paved road on the mountain's west side. But the woods around the mountain were full of people who had taken gravel roads to their destinations—loggers, photographers, reporters, scientists, law enforcement personnel, and sightseers who just wanted to be near an erupting volcano.

The group of public officials and industry representatives who had met in Vancouver on March 26 had continued to meet since then, and that group took on the task of devising an access system for the mountain. Within the Forest Service, the job fell primarily to Ed Osmond, a twenty-three-year veteran of the agency who had transferred just a few months earlier to the Gifford Pinchot National Forest from a national forest in Oregon. With a small group of colleagues, Osmond took a set of Forest Service maps and started drawing lines around the volcano. They began by drawing a line across the first ridgeline north of the mountain, just above the north fork of the Toutle River and Spirit Lake. The line then turned north and extended through a high point of land known as Mount Margaret. As Osmond said later, "We did not think anything would go beyond Mount Margaret ridge in any way, shape, or form. We thought any type of event would be contained in the physiographic basin." The land through which this line passed was within the Gifford Pinchot National Forest, though Weyerhaeuser, Burlington Northern, and a few other companies still owned scattered square-mile sections inside the national forest. But most of their land was north of the line, so they wouldn't complain.

To the east of the volcano, Osmond and his colleagues carried the line down a ridge between Smith Creek and Bean Creek, figuring, as on the north side, that the ridgeline would block the force of any eruption. On the south, the line followed a road that provided access to some of the southern climbing routes to the mountain. It was a bit close to the volcano, but putting it any farther south would place off limits some of the rich forestlands being harvested by the Forest Service and private companies.

The problem was on the volcano's western and northwestern sides. On those sides, Weyerhaeuser owned almost all the land outside the national forest. But the boundary between the national forest and Weyerhaeuser land, which ran north-south along a county line just to the west of the mountain, was less than three miles from the peak. Osmond and his colleagues felt that they could not draw their line onto Weyerhaeuser property, because they did not think they could force Weyerhaeuser to remove employees from its own land. So on the western and northwestern sides, they simply followed the boundary between the national forest and Weyerhaeuser's property.

Everything inside this line, Osmond and his associates called the red zone. Only scientists and law enforcement personnel were allowed into this area. True, Harry Truman was holding out at his lodge on Spirit Lake, which was solidly in the red zone, but he would have to be an exception.

Outside the red zone, Osmond and his colleagues drew another line that generally followed roads ten or so miles away from the volcano's peak. The area between the two lines they called the blue zone. Here loggers and property owners could enter during the day if they had permission, which generally meant acknowledging the risks they were taking. But a problem again arose on the western and northwestern sides of the volcano. Extending the blue zone into Weyerhaeuser property would mean extending the Forest Service's jurisdiction onto private land, and Osmond felt that they couldn't do that. Therefore, on the western and northwestern sides of the volcano, there was no blue zone.

It made for an odd-looking hazards map. On the northeastern, eastern, and southern sides of the mountain, the red and blue zones extended out about five and ten miles, respectively, from the smoldering volcano. But on the western and northwestern sides, the danger zone was curiously squashed. The map looked like a pie out of which someone had cut a large slice. And by an unfortunate coincidence, the bulge on the volcano was pointing almost directly at the area where the danger zones were thinnest.

The geologists and law enforcement personnel were not happy with the danger zones. As Crandell said later, people "wanted the zone to extend as far out as possible, as long as it didn't include their lands." One state patrol captain put it this way: "If this isn't Weyerhaeuser and county politics, it makes no sense at all."

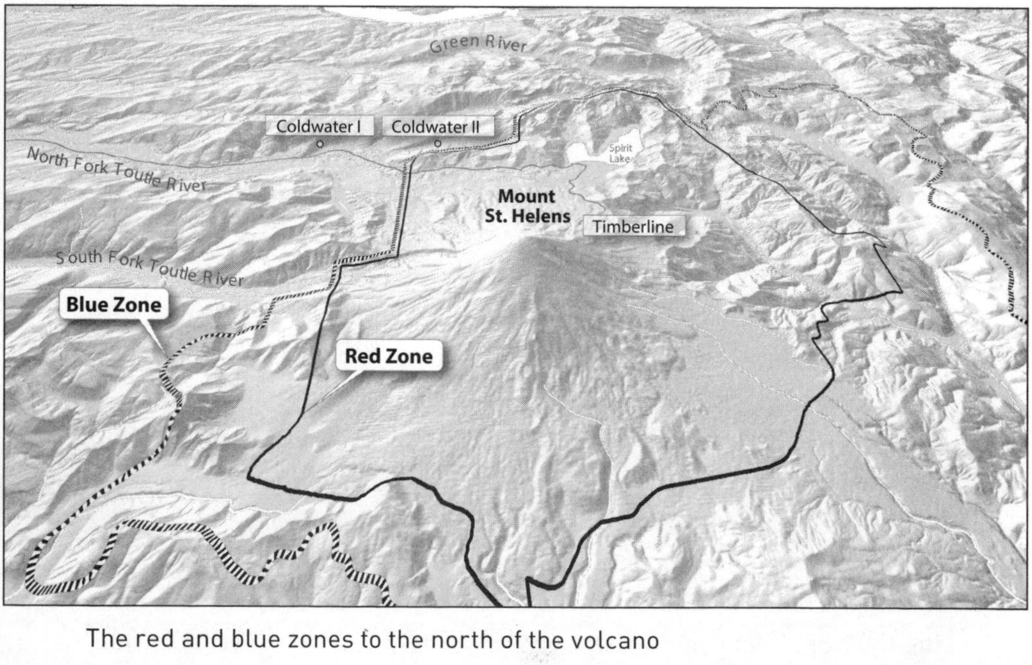

The red and blue zones to the north of the volcano

Other groups besides Weyerhaeuser would have objected if Osmond and the others had drawn the lines for the red and blue zones any farther west. One of the biggest interest groups was the state of Washington itself. Weyerhaeuser owned much but not all of the land between Mount St. Helens and Interstate 5. The state also owned valuable land left over from the land grants, and it had become dependent on that land. The Washington State constitution says that proceeds from state timber sales are to be distributed to local school districts to help pay for construction costs. Timber sales had brought in $175 million for public education in Washington the year before (the equivalent of about a half billion dollars today), and in a state without an income tax, that money was desperately needed. As the head of the state's Department of Natural Resources, Bert Cole, said after the eruption, "We were anxious to continue to harvest timber. . . . We were concerned about the safety of lives, too. But we had timber sales going on in there that had to be taken care of."

Only one person could extend the danger zones into Weyerhaeuser territory—the governor of Washington State, a woman named Dixy Lee Ray. Born Marguerite Ray to a working-class family in Tacoma (her father was a commercial printer), she'd been a tomboy growing up and was often called "the little dickens," which gradually became Dixy. At sixteen she legally changed her name to Dixy Lee Ray, commemorating her tumultuous childhood and a family connection to the Civil War general. Always fascinated by the animals she discovered in the tide pools and waters of Puget Sound, she received a bachelor's degree in zoology from Mills College in Oakland and a PhD from Stanford University in biological science. She worked her way through college as a puppeteer, janitor, waitress, and housepainter. She said, "I missed a lot of sleep in those days, but never a class."

She was a biologist on the faculty of the University of Washington for more than twenty years and then the director of the Pacific Science Center in Seattle, which was built for the 1962 World's Fair and had a profound influence on many science-oriented kids in the Northwest. In 1972, Richard Nixon appointed her to serve on the Atomic Energy Commission, and she became chair of the commission the next year when former chair James Schlesinger was named head of the Central Intelligence Agency. She had always been adamantly pro-nuclear, though her expertise was in marine biology rather than nuclear science. She also had a tendency to say exactly what she thought, even if it wasn't true. For example: "There is no evidence that survivors of the Hiroshima bombings have suffered any more cancer than anyone else, including the second generation. The problems facing the nuclear industry are largely raised by fears of the public, but we all know that fear requires ignorance." She made headlines in Washington, DC, for living in a motor home in rural Virginia, wearing white knee socks, and bringing her Scottish deerhound and miniature poodle to work.

In 1975 she returned to Washington State and decided to run for governor. Never much interested in party politics, she filed as a Democrat since the Republican slate was already full. It was the election after the Watergate scandal, and Ray swept into office with many other newcomers and outsiders, including Jimmy Carter. When asked on election night to explain her success, she replied, "It can't be because I'm so pretty?"

During her time as governor, Ray lived in the governor's mansion with her dogs, although she tried to return to her mobile home on nearby Fox Island on weekends. She quickly drew criticism from her new political party for her conservative views, especially on environmental and energy issues. She sought to permit oil supertankers to offload in Puget Sound but was blocked by Sen. Warren Magnuson, sparking a feud that continued for years. She also lost favor with many women supporters when she refused to campaign for the Equal Rights Amendment to the United States Constitution. Quick to anger, she engaged in legendary feuds with the news media. In 1978 she named a litter of eleven piglets born at her Fox Island home after various reporters. The next year she treated reporters to sausages made from those pigs.

When the first phreatic eruption occurred on Mount St. Helens on March 27, Ray was addressing a meeting of Washington Superior Court judges in Port Ludlow, an old timber town on Puget Sound. "I might just read you the note that has just been handed to me," she told the assembled judges with a grin. "We have received information that Mount St. Helens has erupted at 12:58 today. . . . I've always said, for many years, that I hoped I lived long enough to see one of our volcanoes erupt. Maybe soon I will get a chance." After the speech, she went directly to a nearby airport and had a plane fly her to the mountain, after which she reported, "It was really quite a thrill. . . . No lava, [but] the top is quite dirty."

Toward the end of April, the state Department of Emergency Services forwarded the map prepared by Osmond and his colleagues to Ray with an urgent request from local law enforcement that she establish the red and blue zones in state policy. The map was obviously deficient. Ray knew that people were using logging roads to get way too close to the mountain on the western and northwestern sides. But she, like Osmond, did not want to extend the red or blue zones onto Weyerhaeuser property. "You cannot restrict or remove people from their homes or prevent them from earning a living unless you have awfully good reason," she later said. The proposed map's failure to impose any restrictions on Weyerhaeuser's land gave Ray and her aides an excuse not to do so either.

But that decision had major consequences. Without an extension of the red or blue zones to the west or northwest, the people entering those areas were doing nothing wrong and could not be told by law enforcement to leave. Evidently, to Ray, they were on their own and should have known better. "It's just impossible to believe that there could be anybody who didn't realize that coming close to the mountain was hazardous." On April 30, she issued an executive order that simply adopted the red and blue zones developed by the Forest Service. It was an act that would have dire consequences.

• • •

Toward the third week of April, about the same time the public learned about the bulge on Mount St. Helens's north slope, the worst possible thing happened: the mountain began to quiet down. The number of earthquakes and steam and ash eruptions fell, though the intensity of the quakes was greater, so that the total seismic energy shaking the mountain was about the same. But the mountain was strangely still. Only the bulge was active, growing and expanding until the north face of the mountain became grotesquely bloated and cracked.

That same week, on April 24, the US military had launched a rescue attempt of fifty-three hostages in Iran that went terribly wrong. Of eight helicopters that took off from the *Nimitz* aircraft carrier off the coast of Iran, only six made it to a staging area in the desert four hundred miles southeast of Tehran, and another one had a mechanical failure when it arrived. In a move still debated in military circles today, President Carter agreed with the commander that the mission should be aborted. Then, as one of the helicopters was being moved for refueling, it hit a troop plane, causing an explosion and fire that killed eight servicemen. Five helicopters were left behind in the retreat, some of which contained the names of Iranians working for the Americans. The debacle sapped US prestige and contributed to the outcome of that fall's presidential election.

The press, distracted by the hostage crisis and the election, began to lose interest in Mount St. Helens. The press conferences in Vancouver fell from three a day to twice a week, and by the beginning of May they ended altogether. The geologists would record a message onto a telephone answering machine, and any reporters writing stories would call in to learn what was happening.

As the mountain got quieter, several groups affected by the red-zone restrictions got louder. The loudest was the group of people who owned homes and operated resorts around Spirit Lake and the north fork of the Toutle River. The roadblock on the Spirit Lake Highway seemed designed only for them. Loggers were still working on the ridges right next to Spirit Lake, and the hills were filled with photographers, geologists, and sightseers, but the property owners who had the most at stake in the land couldn't go to their homes. In Washington State, property tax bills are sent out in April, and the arrival of tax bills in the mail raised new howls of protest. The state wouldn't let property owners near their homes but it still expected them to pay taxes on those homes. The property owners knew that snow avalanches in the winter rarely got past the Timberline parking lot, and they certainly never reached Spirit Lake. Why should a rock avalanche be any worse?

Some of the loggers working for Weyerhaeuser also were agitated, but for the opposite reason. On the rare occasions when the weather cleared and they saw the bulge on the mountain, they knew they were working in a dangerous spot. Maybe the rocks distended by the bulge would drop harmlessly into the Toutle River valley, but that was the road they took to get home every night. Weyerhaeuser had written an emergency plan that kept them up on the ridgelines, along with the sightseers and scientists. But they knew that those ridgelines had been covered with ash in the past, and they didn't want to be there when it happened again. Some of the loggers were working on the ridgeline just to the north of the mountain along which the red-zone boundary ran. If the red zone was too dangerous for members of the public, why wasn't it too dangerous for them?

Washington State has a Division of Industrial Safety and Health, and in April complaints from loggers began to arrive in the Longview office. Senior safety inspector Les Ludwig called several logging companies working around the mountain and asked them to describe the precautions they were taking and their evacuation plans. Weyerhaeuser's response referred to a contingency plan it had developed that called for its loggers to make their way toward high ground in the event of an eruption. For Ludwig, that was enough. "It wasn't my jurisdiction," he later told the *Oregonian* newspaper. "You don't go in and tell a company like Weyerhaeuser not to go in and work. They could have been in there for months and nothing would have happened. Then they would have come back to us with a bad taste in their mouth." The complaints seemed to be coming from just a few men, not large groups. "I thought that most of the complaints were from people who wanted a way to get out of work," Ludwig said. "Frankly, I thought they were out to get unemployment [compensation]. Now, I wouldn't want this to get out in the press, but that's what I thought."

One of the crews working on the ridge north of the mountain was under Jim Pluard, a longtime Weyerhaeuser foreman. His crew was working just on the far side of the ridge north of the mountain, and they weren't happy about it. The ground was blackened by ash, and earthquakes were a constant reminder of how close they were to the volcano. But Pluard was devoted to the company and did not relay his crew's complaints up the line of command. Finally, the men went to Pluard's boss, a man named Gordon McVey, the head cutting-crew foreman for the region. McVey said that they were working on the other side of the red-zone boundary established by the state, so they should be protected from any eruption. They could take a leave of absence without pay if they thought they were in danger, but the assignment would not be changed. "The job is up there if you want it," he told them.

Reluctantly, they kept working. "It was in an area where we shouldn't have been—there's no question about that," one logger said. "We were all shaky-legged about it. But it was a job, and I guess we took a long shot along with the rest of them. I didn't like it up there. We just took a long shot and tried to make a dollar."

By the beginning of May, the US Geological Survey was beginning to take seriously Barry Voight's warning that the Timberline parking lot was a very dangerous place to be working. But the Coldwater observation post where geologists had been monitoring the mountain seemed too far away to provide a timely warning. Early in May, Weyerhaeuser bulldozers removed the snow from a logging road on a ridge just north of the mountain, near the slope where Pluard's men were logging. A clearcut from the top of the ridge, just five and a half miles from the summit, provided an unobstructed view of the volcano and its bulging north flank. To this clearcut the geologists hauled a small white trailer, equipment for monitoring the mountain, a radio, and some director's chairs. They named it Coldwater II, after which the original observation post became Coldwater I. Coldwater II was just a few yards from the red-zone line running along the top of the ridge. But a colleague of Rocky Crandell's had done an analysis of the ridge and had concluded that, except for airborne ash deposits, it had been untouched by every eruption in the last 38,000 years.

But at least some of the geologists knew it was not a safe spot. At the beginning of May, one of them contacted the Washington State National Guard to ask if the survey could borrow an M113 armored personnel carrier—essentially, a tank on Caterpillar treads—that would be placed next to the trailer. A note accompanying the order said that "in case of violent explosion, winds (with debris) could reach 100 mph."

The request for the armored personnel carrier never appeared in the newspapers, and even some of the geologists monitoring the mountain didn't hear about it right away. But when they did, they couldn't help but wonder. If whoever was monitoring the volcano from Coldwater II needed a tank for protection, what did they think was going to happen up there?

THE KILLIANS PLAN TO GO CAMPING

ON THURSDAY, MAY 15, 1980, ANYONE STOPPING FOR A CUP OF COFfee at the Little Crane Café would almost certainly have leafed through a copy of the *Longview Daily News* left lying on the counter. There on the front page of the B section was a picture of Harry Truman at an elementary school in Oregon. On Wednesday he'd been helicoptered to the school by *National Geographic* for a story. When he landed, all 104 of the school's students stood outside holding a banner that said HARRY—WE LOVE YOU! He told them about his half century at the lodge and why he would never leave. When one of the students at the school asked if he knew when the volcano would erupt, he replied, "I wish I did, because I would run. I'm going to tear down that hill as fast as I can."

Next to that was a story by Donna duBeth, who had been doing much of the volcano coverage for the newspaper. Under the headline "Quakes, Bulges, Ash Continue," the story began:

> Earthquakes, eruptions, continued bulging. Mount St. Helens seems locked into that pattern and so do the geologists who are keeping track of the mainland's only active volcano.
>
> "We are seeing eruptions of steam and ash similar to those of

> the past few days," the U.S. Geological Survey reported this morning.
>
> That statement duplicates information heard most mornings by reporters calling the Vancouver headquarters for the latest mountain news.

Nothing had changed. Except for the occasional earthquake, the mountain was quiet. The bulge was continuing to grow, but no one knew if or when it might give way. The long wait for something to happen was becoming tedious.

John and Christy Killian were making plans that Thursday to go camping over the weekend at Fawn Lake, in the high country between the Green River and the north fork of the Toutle. The news item that would most have interested them was farther back in the B section of the newspaper:

> Extended forecast: Partly cloudy with little or no precipitation Saturday through Monday. Near normal temperatures with highs in the 60s. Lows in the 40s.

It would be perfect. The woods would be clear and cold. The lake would be full of fish. He and Christy could get away from their families, just the two of them, alone in a tent, nothing but fishing and eating and sleeping, the clear sky overhead, the smell of woodsmoke, the rippling lake. They would come back from a weekend like that ready to face anything.

PART 3

THE CONSERVATIONISTS

Mount St. Helens rising above Spirit Lake

THE FATHER OF THE FOREST SERVICE

ON AUGUST 22, 1896, WHILE RIDING ON THE NORTHERN PACIFIC line from Tacoma to Kalama, Gifford Pinchot gazed out the window of his Pullman coach at the vast forest that would someday bear his name. It was a magnificent sight. The Douglas fir trees swooped down from the high country like an immense green cloak draped over the land's bare shoulders. Even from a distance, Pinchot could make out the individual trunks, ten or more feet across, standing separately above the undergrowth, like solitary men, proud and independent. The forest reminded him of other great forests he'd seen: the redwoods near Yosemite, the vast pineries of northern Europe, the hardwood forests near the Biltmore Estate in North Carolina. Yet this forest seemed to extend much farther than any of those forests, rising in successive waves to the east, to the distant volcano that Meriwether Lewis, when he was traveling up the Columbia River almost a century earlier, called "the most noble looking object of its kind in nature."

Though he had just turned thirty-one in the summer of 1896, Gifford Pinchot always seemed older than his age. Partly it was his great height and cadaverous bearing, which reminded some onlookers of the trees he so admired. Partly it was the bushy mustache overhanging his upper lip, which he had grown as a student at Yale and had cultivated ever since. And he was a serious man, committed to his calling and to his god, so that his passions were rarely checked by self-doubt. He had wanted to be a forester since he was a senior in high school. Now he was one, and through careful maneuvering he had put himself in a position to make decisions about some of the greatest forests the world had ever seen.

Pinchot always acted with the assurance of someone who expected to do great things. His grandfather, Cyrille Pinchot, had emigrated to America from France in 1816 when his family cast its lot with Napoleon right before Waterloo. Settling first in New York City and then on four hundred acres of rich farmland outside Milford, Pennsylvania, near the Delaware River, the family established a store that exchanged raw materials from western New Jersey, Pennsylvania, and New York for finished goods moving in the opposite direction. As a young man, Cyrille began buying land in the area and hiring tenant farmers to raise crops. He also bought forested land, clearcut the trees, and bound the wood in rafts to float down the Delaware to sell to lumbermen in Trenton, Philadelphia, and other cities. He reinvested his money in more farms, more timberland, and, eventually, property in the expanding towns of the region.

By the time Cyrille's second son, James, came of age in the 1850s, the family fortune was sufficient for a pivot away from farming, timber, and country stores. James moved to New York and started a lucrative business in interior furnishings—wallpaper, window shades, and curtains—while also moving into the high echelons of New York society. In 1864 he married Mary Jane Eno, the daughter of a wealthy New York merchant and land speculator. They were friends and patrons of several prominent landscape artists, including Sanford Gifford, after whom James and Mary would name their first son in 1865. James became involved with the National Academy of Design and helped establish the American Museum of Natural History. He was in the vanguard of the city's social, economic, and cultural elite.

Gifford Pinchot grew up in New York City, spent his summers in Simsbury, Connecticut, and attended Phillips Exeter Academy in New Hampshire before entering Yale in 1885. He was groomed for a life of service, not commerce. When he was growing up, his parents gave him a copy of a book that was hugely influential in the second half of the nineteenth century, *Man and Nature* by George Perkins Marsh. In that book, published in 1864, Marsh—a scholar of ancient languages, a former congressman from Vermont, and a longtime diplomat in Italy—argued that the classical civilizations bordering the Mediterranean had collapsed because of environmental degradation. In particular, by cutting their forests, these civilizations had impoverished the soil and hastened the spread of deserts, which weakened the foundations of government and society. Now Americans were embarked upon the same heedless course. By destroying the forests of the eastern United States and Midwest, the country was destined to collapse just like Greece and Rome, Marsh wrote. Only if the nation restored its land, conserved its forests, and protected its soils and waterways would it avoid catastrophic decline.

Gifford Pinchot's father had always regretted the desolate condition in which his family had left the woodlands of eastern Pennsylvania during its rise to prominence and wealth. For him, the ideal landscapes were the pastorals depicted in Sanford Gifford's paintings; the second-growth forests of the Adirondacks, where the family went on camping and fishing trips; or the cultivated fields and copses seen through railway windows on European sojourns. In his son, James Pinchot saw a chance to rectify his father's and grandfather's mistakes. Right before Gifford entered Yale, his father asked him, "How would you like to be a forester?" It was "an amazing question for that day and generation," Pinchot wrote later in his autobiography. In 1885, no American university offered courses in the establishment, management, and preservation of forests, and no American had ever become a forester, though some European-trained foresters were working in the United States. The idea that an American would make forestry a career, in an age when the forests were being rapidly exterminated from large swaths of the country, seemed a non sequitur.

Pinchot was immediately enamored of the idea. Throughout his time at Yale, he proclaimed his intentions to his classmates and pestered his science professors for information about his chosen profession, even though there was little they could tell him. He also threw himself into Yale life with abandon. He was a reserve on the great Walter Camp football teams of the 1880s, taught in the Yale Sunday school, and was tapped for Skull and Bones at the end of his junior year. During his senior year, he flirted with the idea of going to work for the YMCA at Yale after graduation—both of his parents had also instilled in him a powerful religious sense—but in the end he opted for forestry, making as a graduation speaker his "first public statement on the importance of Forestry to the United States."

For several months he traveled through Europe gathering forestry texts, meeting with prominent foresters, and touring the meticulously managed forests of England, France, and Germany. At the 1889 World's Fair in Paris, Pinchot was more impressed by the *Eaux et Forêts* exhibit than by the newly built Eiffel Tower. The exhibit provided an exhaustive survey of forestry and hydrology, extolling the rational management of the landscape and the triumph of French forestry over nature. "The exhibit is magnificent and so complete that at first I despaired of making any adequate description of it," Pinchot wrote in his dairy. That fall he enrolled at L'École Nationale Forestrière in Nancy, intending to make a thorough study of the subject. But he preferred hiking in the nearby woods to his classes, where he observed how the forests were being cut, reforested, and then cut again, all through the application of sound scientific principles by civic-minded foresters. In this way, he wrote, "a permanent population of trained men" could enable "permanent forest industries, supported and guaranteed by a fixed and annual supply of trees ready for the ax."

Though the European foresters he met encouraged him to remain in Europe and earn a PhD, Pinchot was impatient to get started on his career. In December, he set sail for America, "willing to try with what knowledge I have now." He wrote for outdoors periodicals and spoke at national conferences on issues related to forestry, met with the European-educated foresters then working in America, and did some work for Phelps Dodge & Company, a large mining and timber concern with which he had family connections. In 1892 he took a job as the forester of George Vanderbilt's Biltmore Estate in North Carolina, where he first applied his ideas about how to manage American forests. He was becoming known in the small community of university and government scientists interested in forest preservation, and no one was surprised when, in the spring of 1896, he was named secretary of the National Academy of Sciences' Committee on the Inauguration of a Rational Forest Policy for the Forested Lands of the United States—more popularly known as the National Forest Commission.

The impetus to establish such a commission had been growing for years. The three decades after the Civil War were a period of steadily increasing environmental concern in the United States, triggered largely by the industrial cutting of America's forests. Marsh's book was one of several that predicted disaster if the despoliation of the land continued. Writers raised concerns about a "timber famine," such as the widespread shortages that had been occurring in parts of western Europe for centuries. As the secretary of the Interior warned in 1877, "If we go on at the present rate, the supply of timber in the United States will, in less than twenty years, fall considerably short of our home necessities." Despite such warnings, the destruction of forests in New England, the Mid-Atlantic states, and large parts of the South during the United States' first century was immediately followed by the widespread devastation of forests in Michigan, Wisconsin, and Minnesota at the beginning of its second. Once the trees were gone, the leftover slash was often burned, leaving a wasteland of charred stumps. What to do with this cutover land was a major problem. Often the timber companies simply quit paying local taxes on the property they owned so that the worthless land reverted back to the state. In other cases, the timber companies tried to sell the cutover land to farmers, but the sandy soils and harsh winters of the upper Midwest reduced many such farmers to penury and then bankruptcy. The eminent forest historian Michael Williams describes the scene:

> There was a high rate of failure, and the survivors hung on to lead a wretched life, trying to eke out an existence. The cutover areas were (and still are in places) dotted with derelict fences and sagging, unpainted farmhouses, some mere tar-paper shacks. In the

> deserted fields occasionally one still sees a lilac tree or a heap of stones where a chimney once stood, both markers of an abandoned homestead, the whole scene a mute and melancholy testimony to abandoned hopes in the former forested lands.

Even as the forests of the Midwest were being eliminated, a new generation of advocates like Walt Whitman and John Muir, building on the transcendentalism of Thoreau and Emerson, were defending forests for their own sake, as mystical places where individuals could re-create themselves, realizing the inherent worth and dignity of both people and nature. American landscape painters depicted a world in which humans and nature coexisted, each acquiring attributes of the other in a romantic union of man and his environment. In a time of religious revival, the destruction of the forests seemed to represent the expulsion of Americans from their Edenic paradise. Only the preservation of wilderness could safeguard America's innocence.

In 1891, concerns about American forests coalesced in a remarkable piece of legislation, which Pinchot later called "the most important legislation in the history of Forestry in America." Without quite knowing what it had done, Congress passed a law that included an obscure provision stipulating that "the President of the United States may, from time to time, set apart and reserve [land] in any State or territory having public land bearing forests . . . as public reservations." Almost immediately, President Benjamin Harrison created 13 million acres of reserves from Arizona to Washington. His successor, Grover Cleveland, added another 4.5 million acres a year later.

The 1891 legislation marked a dramatic turn in America's policies toward its vast landholdings. Except for areas of exceptional natural beauty—Yellowstone National Park was created in 1872 as "a public park or pleasuring ground for the benefit and enjoyment of the people," followed by the creation of Yosemite National Park in 1890—government policy had always been oriented toward "disposal"—getting land into the hands of private owners as quickly as possible. Yeoman farmers were the preferred owners, stemming from the Jeffersonian belief that small landowners were the best bulwarks against tyranny. Beginning during the Civil War, homestead acts granted quarter-square-mile plots to claimants willing to settle the land and farm.

By the 1890s it was clear that these policies had gone seriously awry in the western United States. Farmers were largely uninterested in the vast majority of the land west of the 100th meridian, where rainfall was too sparse to support crops. Even the great forests of the West generally grew on land too steep, too isolated, or too rocky for agriculture. Instead, timber companies sought to acquire this land, whether legally or illegally. Many companies paid loggers, sailors, and other surrogates to claim forested lands as if they were going to clear the trees and raise crops, after which the claimants signed over the deeds to the companies. In many cases, western settlers simply assumed that the land was theirs to use, regardless of ownership. Timber was brazenly stolen from federally owned lands by settlers, mining prospectors, railroad contractors, and others, with no payment to the government. Ranchers fenced in vast tracts of public land to graze their livestock. The western lands were being lost or given away without compensation. As a land commissioner wrote in 1891, "A national calamity is being rapidly and surely brought upon the country by the useless destruction of the forests. Much of this destruction arises from the abuses of the beneficent laws for giving lands to the landless."

The creation of the first forest reserves in 1891 heralded a new era. Henceforth, the government would set aside large portions of the American West—and eventually portions of eastern forests—as federally owned land reserved for the use of the American people as a whole. No longer was the intention to give away or sell every last piece of land. Some of it would be reserved for the greater good. What Wallace Stegner said of the national parks applied to the forest reserves as well: They were "the best idea we ever had. Absolutely American, absolutely democratic, they reflect us at our best rather than our worst. Without them, millions of American lives, including mine, would have been poorer."

Though many westerners supported the creation of the forest reserves, their establishment also caused great consternation among politicians, landowners, and businessmen—especially those who had been getting rich from the exploitation of western lands. The bill sanctioning the reserves contained no provisions for their management or use. How was the federal government to manage these lands? Would the land still be open to logging, mining, and grazing? Would the reserves be expanded into other areas? Were they forever to be off limits, as their name would suggest?

To answer these questions, the secretary of the Interior requested that the National Academy of Sciences in Washington, DC, form the National Forest Commission to investigate whether it was "desirable and practicable to preserve from fire and to maintain permanently as forested lands those portions of the public domain now bearing wood." The chairman of the commission was Charles Sprague Sargent, the curator of the Arnold Arboretum at Harvard and author of the monumental 1884 *Report on the Forests of North America.* The commission also included John Muir, who had founded the Sierra Club in 1892 and insisted on being an "observer" to the commission to maintain his independence from its conclusions; Henry Abbott, an army engineer; William Brewer, a prominent Harvard botanist; Arnold Hague of the US Geological Survey; and the zoologist and engineer Alexander Agassiz. And thirty-year-old Gifford Pinchot, the only member of the commission who was not a member of the academy, was named its secretary.

Of course, the commissioners could not very well recommend what to do with land they had never seen, so in the summer of 1896 they set off on a grand tour of the American West. Meeting in western Montana, they hiked through the Bitterroot Valley and nearby portions of Montana and Idaho. Traveling on the Northern Pacific line, the commissioners passed through Spokane, over Stampede Pass to Tacoma, and then south to Portland. By coach, railcar, and flatboat, they journeyed along the Willamette, Rogue, and Umpqua valleys of Oregon, visiting Klamath and Crater Lakes. Then they traveled south to California, where they viewed the Sierra, San Bernardino, and San Jacinto mountain ranges. They went to Arizona, touring the Grand Canyon, and passed through New Mexico on their way to Pikes Peak, where they rode to the top in a stagecoach. By October they were heading back to Washington, DC, to write their report.

Even during their trip through the West, divisions on the commission were becoming apparent—divisions that continue to split environmentalists to this day. Muir and Sargent, joined by Agassiz and Abbott, were preservationists. They wanted not only to stop the destruction of the forests but to halt their future development. President Cleveland should immediately create more forest reserves, they believed, and then use the army to guard the reserves from timber thieves and fire.

Pinchot, joined by Hague, had a very different opinion. Pinchot's study of forestry in Europe had convinced him that the forests could be used without being destroyed. He favored the utilitarian conservation of resources, not their inviolate preservation. The application of science through an enlightened corps of well-trained foresters could provide all things to all people—timber, water, recreation, and wildlife preservation—an idea that eventually became known as the "multiple use" doctrine. Pinchot wanted the commission to urge Congress to develop a management plan for the forests before Cleveland created any additional forest reserves. That way, Congress would be less likely to object to the creation of future reserves and roll back the progress already made.

Pinchot lost this battle. In a preliminary report to the secretary of the Interior on February 1, 1897, the commission recommended the establishment of thirteen new forest reserves covering more than 21 million acres, to be added to the seventeen reserves already existing. The report called for the army to protect the reserves and made no mention of how they were to be managed or used. Three weeks later, on February 22, President Cleveland issued proclamations establishing the reserves recommended by the commission, just two weeks before he was to leave the White House. The act brought him great satisfaction. When three officials from the American Forestry Association, which was founded in 1875 during the United States' first wave of environmental activism, came to thank him, they were ushered into the Oval Office past a row of politicians who had been waiting for an audience. "This is the first time anybody has thanked me for anything for a long time," Cleveland said. "Let that bunch of favor-seekers rot in their chairs." They talked for more than an hour, during which Cleveland exclaimed that he had created twice as many reservations as most men would have thought advisable. "A Republican president will succeed me, and he will undo half of what I have done, so to be safe I have done aplenty."

• • •

By 1905, Pinchot had regrouped from the National Forest Commission's rejection of his ideas. He was ready to try again to place America's forests under sound management principles. And this time he succeeded.

Seven years earlier, on the basis of his work on the National Forest Commission and his many connections in government, Pinchot had been named Chief of the Division of Forestry in the US Department of Agriculture. The title was soon changed to "Forester," which pleased Pinchot, since "in Washington chiefs of division were thick as leaves in Vallombrosa. Foresters were not." It was a grand title for a job with little real power. The forest reserves were under the Department of the Interior, which also oversaw the national parks. The Division of Forestry in the Agriculture Department had no jurisdiction over the reserves or any other public lands. As Charles Sargent, the former president of the Forest Commission, wrote to John Muir, "This is a good place for [Pinchot]. He can do no harm there and after a very short time people will cease to pay any attention to what he says."

Pinchot, as usual, proposed a big solution for what he saw as a big problem. He began to advocate for moving the forest reserves from the Department of the Interior to the Department of Agriculture. There the forests could be treated not as an untouchable relic, like the national parks, but as an agricultural commodity. They could produce, as he put it in his autobiography, "the greatest good, for the greatest number, for the longest run."

Moving that much authority from one department to another is not an easy thing to do in Washington, DC, where the size of a department's budget and clout is measured by the programs under its control. Success would take every bit of Pinchot's persuasive power.

His cause was aided immeasurably by an assassin's bullet. On September 6, 1901, an anarchist who had lost his job in the panic of 1893 shot President William McKinley as he was shaking hands with members of the public at the Pan-American Exposition in Buffalo, New

York. McKinley died of gangrene eight days later after surgeons were unable to locate the bullet that had entered his abdomen. This brought to the presidency a man with whom Pinchot already had an intimate friendship: Teddy Roosevelt. Just two years earlier, when Roosevelt was governor of New York, he and Pinchot had engaged in a round of wrestling in the executive mansion in Albany—won by Roosevelt—followed by a round of boxing, which Pinchot won. "I had the honor of knocking the future President of the United States off his very solid pins," Pinchot would later boast. When Roosevelt became president, the two men chopped wood for exercise together, hiked through Washington's Rock Creek Park, rode horses, and swam in the Potomac River. Pinchot ghostwrote Roosevelt's speeches on conservation, drafted legislation, and served as one of his closest advisers. Roosevelt was an ardent outdoorsman who burned to protect the natural places he so loved. But he could not have made nearly as much progress without Pinchot.

Pinchot had already been preparing the Agriculture Department to absorb the forest reserves. He had been hiring young men to work as foresters with the states and with private landowners, and many lumbermen, who themselves were worried about running out of wood, gladly accepted Pinchot's offers of help. Many of the new foresters had degrees from the Yale School of Forestry, launched in the fall of 1900 with a $150,000 endowment (soon doubled) from the Pinchot family. Amazingly, Pinchot had even convinced the secretary of the Interior to back the transfer of the forest reserves, and Roosevelt also supported the shift of power to his friend. But the only way to transfer the forest reserves would be through congressional action, which would be very difficult to obtain.

Pinchot planned a major event as the culmination of his campaign. The American Forest Congress, held January 2–6, 1905, brought together four hundred representatives of all the groups with a stake in forest preservation, including prominent members of the lumber industry, railroads, grazing, irrigation, and government. Railroad baron James J. Hill, still touting his western holdings, wrote a letter saying that "the subject is of importance far beyond the general understanding of the public. . . . Irrigation and forestry are the two subjects which are to have a greater effect on the future prosperity of the United States than any other public questions." Though Frederick Weyerhaeuser was ill and unable to attend, his youngest son, F. E. Weyerhaeuser, told Congress that "at present lumbermen are ready to consider seriously any proposition which may be made by those who have the conservative use of the forests at heart." Roosevelt opened the conference with a speech, ghostwritten by Pinchot, entitled "The Forest in the Life of a Nation," in which he praised Pinchot for bringing about a "business view" of the forests and posed the question "how best to combine use with preservation." As Pinchot himself said, conservation was simply "good business."

An odd event occurred at the convention. At one point Roosevelt departed from the text Pinchot had written for him and began to denounce timbermen, many of whom were in the audience. He castigated the men who "skin the country and go somewhere else . . . whose idea of developing the country is to cut every stick of timber off of it and then leave a barren desert for the homemaker who comes in after him. That man is a curse and not a blessing to the country." These offhand remarks set back relations between government and the timber industry for years.

Despite this dustup, Congress was convinced. On February 1, 1905, it transferred the forest reserves from the Department of the Interior to the Department of Agriculture, where they have remained to this day. The forest reserves soon were renamed the national forests, in keeping with Pinchot's idea that they were there to be used, not reserved, and the Bureau of Forestry in the Interior Department became the Forest Service in the Agriculture Department. When Roosevelt became president in 1901, fifty-four forest reserves covered about 85 million acres. By the time he left office, the total area included in the national forests was more than 150 million acres, an area larger than California and New York combined. Over the course of Roosevelt's presidency, he and Pinchot did more to preserve America's forests than anyone else in US history. By that measure at least, Roosevelt earned his spot on Mount Rushmore with Washington, Jefferson, and Lincoln.

Frederick Weyerhaeuser (left), who built a lumber empire from a small sawmill on the banks of the Mississippi River, and James J. Hill (right), his next-door neighbor in St. Paul

Gifford Pinchot (above), at about the time he served as secretary for the National Forest Commission, and President Theodore Roosevelt (right)

Painting by Canadian artist Paul Kane of the 1847 eruption of Mount St. Helens

Mount St. Helens and Spirit Lake before the 1980 eruption

The newly formed crater on top of the volcano in late March 1980

Geologist Dave Johnston collecting samples from the crater in April 1980

The Coldwater I station on the edge of a ridge eight miles from the volcano

Dave Johnston at Coldwater II on May 17

An ash-covered Mount St. Helens from Coldwater II on May 17

Mount St. Helens erupting on Sunday, May 18, 1980

The blast cloud on the east side of the volcano

Fawn Lake after the eruption, where John and Christy Killian were camping

A Weyerhaeuser company crummy submerged in a mudflow from the volcano

Rescuers looking at the remains of photographer Reid Blackburn in his ash-filled car at Coldwater I

Ralph Killian looking for his son at Fawn Lake the year after the eruption

The Green River valley looking west toward the area where Clyde Croft and Al Handy were camping on the morning of the eruption

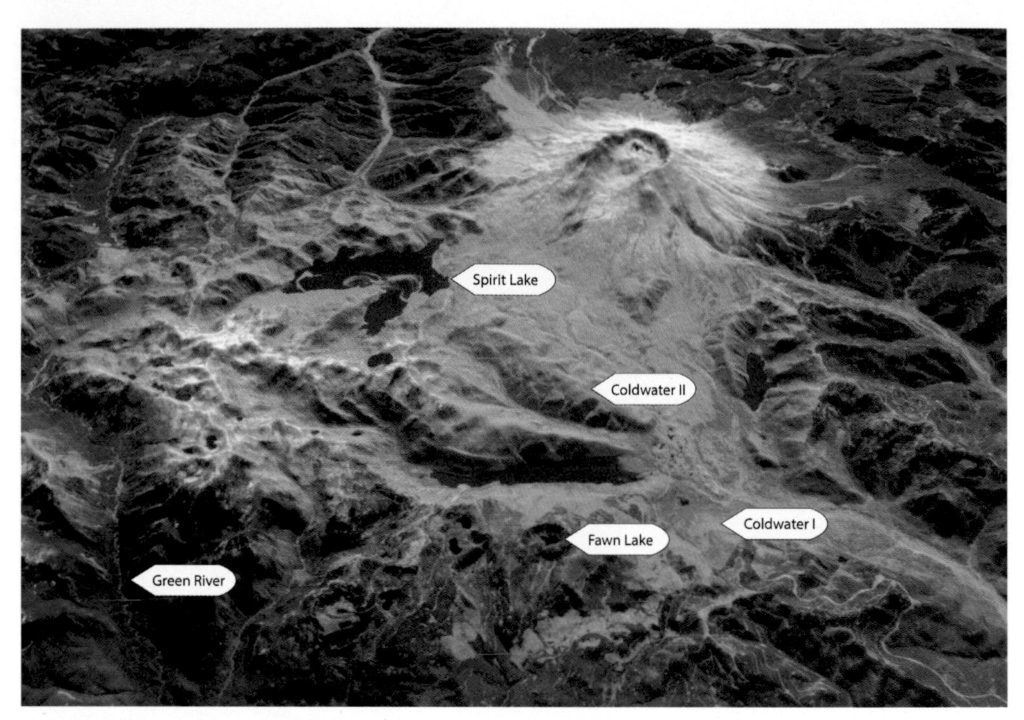

Satellite photograph of Mount St. Helens taken in the spring of 2015

Mount St. Helens in 1982, with a new volcanic cone rising from its crater

When President Cleveland created the Mount Rainier Forest Reserve in 1897, Mount St. Helens was left out of the reserve, partly because of the commercial value of the forests right around the mountain. But an immense 1902 forest fire just south of the volcano dimmed private companies' enthusiasm for the area. Five years later, Roosevelt and Pinchot succeeded in adding the area around the volcano to the renamed Columbia National Forest. That pushed the boundary of the national forest right up against Weyerhaeuser's vast forestlands to the west.

That boundary is where Ed Osmond and his colleagues drew the lines for the red and blue zones in April 1980.

• • •

Gifford Pinchot, who was born in the final year of the Civil War, died the year after World War II ended. After a vigorous campaign to rename one of the national forests for the father of the Forest Service, Columbia National Forest was chosen because it was "one of the top forests in the United States." On June 15, 1949, President Harry Truman signed the proclamation designating the Columbia National Forest the Gifford Pinchot National Forest. At the dedication ceremony later that year, Lyle Watts, the chief forester of the US Forest Service, said that those who worked with Pinchot remembered him "as a man of tremendous energy and enthusiasm, as an inspiring leader, as a zealous crusader. They knew him as a courageous, unflinching fighter in the public interest and for the public good. His cause did not need to be popular if it was right. He seemed at times to be fighting almost single-handedly; but history has shown that the people were behind him."

Today, Gifford Pinchot is an ambiguous figure among environmentalists. People respect him for his vision and accomplishments. But by the time of his death, the shortcomings of his "multiple use" philosophy already were apparent. Before World War II, the Forest Service was largely a caretaker organization. Most of the nation's lumber came from private land, and the Great Depression and war limited the demand for wood. That all changed after the war. As had been predicted for decades, private landowners were running out of timber. Most of the big companies like Weyerhaeuser still had uncut land, but small-scale loggers needed stumpage. Now, the loggers said, was the time to put Gifford Pinchot's multiple-use philosophy to work.

The Forest Service had little trouble transitioning to this new era of "getting the cut out." Returning servicemen studied forestry under the GI Bill, and many of these schools were oriented around the needs of the timber industry. The rapidly expanding suburbs demanded huge quantities of wood for joists, flooring, sills, studs, headers, millwork, stairs, siding, rafters, and plywood sheathing. At the time, forestry schools taught that old-growth forests were unproductive, disease-ridden, "decadent" fire hazards that needed to be removed so that new timber could grow. Both at the local and national level, producing more wood to boost the economy and help win the Cold War could make the Forest Service a bigger agency. The Forest Service built roads through isolated wilderness so that private logging trucks could bring out wood, becoming in essence the single largest road contractor in the nation. It had employees who did nothing but serve the timber industry and who were promoted and rewarded on the basis of how much timber they could liquidate. Highly efficient chainsaws could cut down forests far faster than men with handheld saws and axes. Track-mounted steel towers could hoist trees to landings, replacing the old practice of erecting hoists on tall spur trees. The cut on national forests in Washington and Oregon increased from 1 billion board feet in 1946 to a high of almost 6 billion board feet in 1973.

The greatly expanded harvests from the national forests served the needs of lumbermen and the economy. But the large-scale harvesting of western forests quickly ran up against a fundamental problem. More and more people wanted to use the forests for purposes other than timber production. Servicemen back from the war took their rapidly growing families into the woods on inexpensive family vacations. Outdoor companies like REI, founded in Seattle in 1938, began to sell lightweight and waterproof clothes and camping equipment, making it easier to camp, hike, and hunt. Hiking clubs like the Mountaineers in Washington State and the Mazamas in Oregon had been active since the end of the nineteenth century, but now the cognoscenti were joined by many more people who had a more casual but enduring connection to the outdoors. People came to the national forests to hike, bike, horseback ride, backpack, hunt, snowmobile, motorcycle, ski, swim, picnic, kayak, fish, camp, and simply drive. Flush with timber money, the Forest Service built extravagant ranger stations staffed with knowledgeable personnel to help people plan their trips. In the minds of many people, the national forests became largely indistinguishable from the national parks, even though the two were managed by different departments to serve different ends.

"Multiple use" was always a misnomer. Many of the functions Pinchot ascribed to forests are inherently in conflict. A clearcut mountainside no longer attracts hikers. Logging reduces the forests to cropland and nothing more. In contrast, preservation of the land generally ruled out logging or mining. Multiple use could mean designating different parts of the forest for different purposes, and for a long time the Forest Service took this approach. But battles over how the forests should be used were inevitable.

One of the most contentious battlegrounds was the Gifford Pinchot National Forest. By the 1970s, ten to twenty square miles of the forest were being cut every year. By 1980, almost all of the old-growth forests more than two hundred years old were gone, with newer forests growing in their stead. The rest of the old growth was fragmented into small, vulnerable blocks, and soon it would be gone. Superb scenery that generations of Washingtonians had admired for decades was marred by clearcutting, which the timber industry insisted was the only logical way to cut Douglas fir trees. Some of the clearcuts next to highways were hidden by screens of trees, which the Forest Service called "beauty strips," so that passing motorists could not witness the destruction, but others were in clear sight on the sides of mountains and along river valleys. Sawmills and plywood mills sprung up to saw the timber from the national forests, and communities took shape around those mills. The Gifford Pinchot National Forest, along with many other national forests, was becoming a shabby patchwork of clearcuts surrounded by second-growth forests, with little original forest left.

Conservationists were increasingly alarmed. Unless they took action, there would be no more forests to conserve. National groups like the Sierra Club, the Wilderness Society, and the Audubon Society stepped up their efforts in the Northwest, and smaller groups took shape to address specific issues. The management of the national forests could no longer be left to the Forest Service, the leaders of these groups contended. A different approach was needed.

THE PROPOSAL

SITTING AT THE KITCHEN TABLE IN HER MODEST LONGVIEW, WASHINGton, home, Susan Saul looked around at the piles of paper surrounding her and tried not to panic. The campaign to save the area around Mount St. Helens was getting nowhere. For decades, groups in Washington and Oregon had been trying to preserve the area around the volcano, but they had made almost no progress. All around the mountain, Weyerhaeuser and other companies were logging right up to the tree line on soils so thin that trees would never grow back. The Forest Service was building roads through the backcountry so loggers could get at the last remaining old growth. Duval was poking holes up and down the Green River valley, and if the company found enough ore, it would convert the valley into a gigantic open-pit copper mine. Soon it would be too late. Much of the area around the mountain would be nothing more than a blasted landscape of cutover forests and mine tailings.

In her kitchen, Saul was surrounded by the tools of the precomputerized community organizer: envelopes, paper, brochures, maps, an IBM Selectric II typewriter, address labels, a waxer, pica rulers, a light table, Wite-Out, correction tape. By the spring of 1980, she was handling most of the direct mailings and lobbying for the Mount St. Helens Protective Association, while her co-chair, Noel McRae, did most of the organization's day-to-day work. She never thought, when she walked into her first meeting three years before, that she'd soon be helping to lead the organization. She just figured that she *should* be involved, given how much hiking she did around the mountain. But the other members of the association had involved her in their work from that very first meeting. They were applying the conservationist Brock Evans's injunction "endless pressure, endlessly applied." They were writing letters, making proposals, meeting with public officials and the press. But they had little to show for their efforts. They needed more people, more pressure.

Sometimes to Saul it seemed hopeless. Look at the town where she lived—Longview, carved out of the primordial forests by Robert Long in the 1920s to house the thousands of millworkers in his sawmills. Even now, in the spring of 1980, despite the housing downturn, the Weyerhaeuser and Longview Fibre plants on the banks of the Columbia River were pumping so much smoke into the air that, when the wind blew from the south, everything was covered with a thin sheen of ash. Any slowdown in the timber industry would put more people out of work, the people who lived on her street, the people she met in the supermarket. And they would blame her and the association.

But the blame lay elsewhere, Saul knew. Weyerhaeuser was about to run out of old-growth trees, after which the company was likely to get out of the old-growth business, since it rarely bought trees from the Forest Service. Saul's father worked in a sawmill in Springfield, Oregon, where she'd grown up. She knew that the industry couldn't survive the way it was going. It was extracting wood from the forests the way a mining company extracted gold from the ground. When the gold ran out, the industry would go belly-up or move elsewhere. It had happened all across the country before. Now it was going to happen in Oregon and Washington too.

So if the forests were going to run out anyway, why not stop cutting a few years earlier? That's all the association wanted—just leave some of the land as it was. The number of people who would enjoy these forests would be far greater than the number who would benefit from cutting every last tree. That would be the greatest good for the largest number.

Saul was an ardent conservationist who had spent most of her life in logging towns. After graduating from the University of Oregon, she had gotten a temporary job with the Fish and Wildlife Service in Malheur County, in the southeast corner of Oregon. Then a permanent job had opened up in Longview. She figured that the hiking would be better near Mount St. Helens, with Mount Adams just to the east and Mount Rainier to the north, and the hiking *was* better, despite all the timber cutting. But Mount St. Helens was the only volcano in the entire state—almost the only volcano in the entire northwest—that wasn't within a national park, a wilderness area, or a recreation area. Unless people like her did something, the area around the mountain would be ruined.

What made the whole thing especially frustrating is that they had to fight not just the timber companies but the Forest Service too. Sometimes it seemed that the Forest Service wouldn't be happy until the entire Gifford Pinchot National Forest was nothing but a sea of stumps. She'd once seen an aerial photograph of the area around Mount St. Helens, and the national forest land east of the mountain was as cutover as the Weyerhaeuser land west of the mountain. The Forest Service was completely opposed to the idea of a monument. A monument would keep the Forest Service from building roads and making timber sales in their land, and they wanted nothing to do with it. "Their land," Saul thought bitterly. The national forests were supposed to belong to the people, not to the Forest Service. But the Forest Service acted as if people were the threat, not the timber and mining companies.

She'd gotten to know Don Bonker, the US representative for Washington's third district. Whenever he came to Longview to give a talk or meet with his constituents, Saul made sure to be there so she could tell him what the association was trying to do. He was sympathetic to what they wanted. He'd supported the land swap between Weyerhaeuser and the Forest Service that saved Miners Creek. But he had other constituents he needed to please too. Whenever she met with him, she got the sense that he just wanted to tell her, *Don't worry. Everything will be fine.* But everything wouldn't be fine without his help.

Lots of people supported the association's goals, at least in principle. In Saul's clippings file, she had the editorials the *Longview Daily*

News had written about their cause, along with all the letters the paper had published when it invited readers to comment on whether roads and logging should be permitted in the Spirit Lake basin. Almost all of the respondents said no, even the loggers and millmen who wrote in. "Just for a few greedy people to cut those beautiful trees: no, no, no," wrote a papermaker in town. "Shut off the timber cutters until they show concern for the public. They could care less," said a civil engineer. "If we let people like George Weyerhaeuser totally devastate this land, we are defeating ourselves," a local carpenter wrote. More and more people were in the woods every year. They knew what they were losing. They weren't going to let Weyerhaeuser and Burlington Northern take it all away from them. The Mount St. Helens Protective Association just needed to get them involved.

Saul and McRae had a plan. They would lead a series of hikes through the national monument the association had proposed. They would show people what was at stake. The first one was planned for Saturday, May 10. They would lead a group up the Green River valley. That's where the conflict was greatest. They'd go from Weyerhaeuser's clearcuts on the lower part of the river to the old-growth forest north of the volcano. It was one of the most magnificent forests left in the entire Northwest, maybe in the entire country. No one could walk among those trees and believe that they should all be cut down. If the association could just get more people into the woods, they could stop Weyerhaeuser and the Forest Service. But time was running out. If they didn't stop the logging soon, there would be nothing left to save.

THE GREEN RIVER HIKE

AT 7:30 IN THE MORNING ON MAY 10, 1980, SAUL DROVE FROM HER home on Nineteenth Street toward the town square in the middle of Longview. She passed the bronze bust of Robert Long in the middle of the square, his stern, bespectacled face gazing east toward the forested hills on the other side of the Cowlitz River, and parked on the north side of the square at the Longview Public Library, a gorgeous Georgian building that Long had built in the 1920s for his employees and their families.

About twenty people were gathered in the library parking lot. Russ Jolley was there, who'd been very active with the Mount St. Helens Protective Association, along with Hermine Soler; Sarah Detherage; Jim Fletcher; Arlene Walker; Harry Deery and his wife, Ruth; Jean Lancaster; Russ Maynard; Mary Ellen Covert; and eight or ten others. Many were members of the Willapa Hills Audubon Society or the Mount St. Helens Hiking Club, people who largely knew what was at stake. But Saul didn't recognize some of the other people in the group, and the association needed new recruits if it was going to make progress.

She and McRae had thought about calling off the hike when Mount St. Helens started acting up in March. But McRae had called the Forest Service, and the person on the phone said that the Green River valley was not in the red or blue zones, so they were free to go there if they wanted. The valley was ten miles and three ridgelines to the north of Mount St. Helens. No one was expecting trouble that far from the volcano.

They divided into separate cars and headed north on Interstate 5 to Castle Rock, where they turned off the interstate onto the Spirit Lake Highway. They followed the highway past Silver Lake, formed just 2,500 years ago by a massive mudflow that dammed the Toutle River, and drove through the town of Toutle. Just past the cluster of stores and campgrounds in Kid Valley, they turned off the Spirit Lake Highway onto Road 2500, a good gravel road that paralleled the Green River north of the Toutle. They were on Weyerhaeuser land now; Road 2500 passed through clearcuts, hillsides newly planted with young trees, and second-growth forests that were almost ready to cut again. A bit more than twenty-five miles from the turnoff, Road 2500 crossed the Green River before heading up the hill toward Fawn Lake, and there they parked in a small clearing on the north side of the bridge. They pulled on rain parkas and wool hats—it was a cloudy and cool day, only about 50 degrees, and though it wasn't raining at the beginning of the hike, the rain could start anytime. Then they headed up Trail 213.

The area right around the parking spot had been cut over and was surrounded by scrubby second-growth trees. But as soon as the group started up the trail they entered the old-growth forest of the Green River valley. Immediately the forest floor around them opened up, so they could see through the trees to a great distance, and the canopy rose far overhead, so that they had to stop walking and look up to see the treetops. Around them rose the massive boles of Douglas firs and western red cedars, trunks so large that five people joining hands could not encircle them. The trees seemed to hold up the roof of the sky, so that walking through them was like being inside an immense building too grand to be fully seen. The color of the trees around them darkened, even as the forest itself lightened, as if the air had acquired a crystalline green hue. The forest contained trees of all ages, seedlings, saplings, supplicants reaching toward the light overhead, and all around them were the decaying trunks of fallen giants, many of them bearing new trees on their rotting surfaces. The variety and abundance of life were everywhere evident: shelf fungi on the sides of trees, moss draped across branches, the beetles and centipedes of the trail, the birds arcing from tree to tree, the squirrels chattering on overhead limbs, elk tracks in the mud, and all the unseen life of the forest, the fish in the river and the animals underground and the decomposers silently returning the fallen logs to the soil. Nothing really dies in such a forest. One living thing blends seamlessly with the rest. The forest itself is alive, its constituent parts just the passing clouds of a summer day.

Many naturalists have compared old-growth forests to churches, but that analogy doesn't quite work in a place like the Green River valley. If a church represents the efforts of the devout to invite God to join them in a place created by people, then a forest is a place where God already resides, and people can choose to recognize or ignore His presence. Saul felt the presence of God in a forest like this just as much as in the Catholic church back in Longview, but in a different way. It was impossible to imagine a forest like this in which God did not exist.

That wasn't the case for the second-growth forests they'd driven through on the way there. Those were the creation of men—genetically honed, even-aged stands planted by humans to meet the needs of humans. Old-growth forests met no needs. They simply were, in a way that bore no questions about purpose or value. They could not be created by men. They could not even be understood by men. They had too many parts that were interconnected in too many ways. Change one part and everything else would change, but in ways that were unpredictable and often inexplicable. This unpredictability removed such forests from the realm of human perspectives and values. The forest did not need to justify or explain itself. It existed outside of instrumental human considerations.

From her father, Saul knew the value of the wood through which she walked. The timber along this river was worth many millions of dollars. To a lumberman, leaving these trees there to grow and die was like walking past a pile of money and leaving it untouched. It made no sense to leave this timber when it could pay for houses and trucks and college educations so children wouldn't have to risk their lives working in these woods. The Green River valley was not easy to reach: a twenty-mile drive up a gravel road from the west, an equally long drive up an equally slow road from the east. Most people would never see this valley in anything other than photographs. They wouldn't have the patience to drive all the way there and hike up a trail with no views or alpine lakes waiting at the end.

But wasn't that the idea—just to have it there, to know that it existed? People did not need to go there for the forest to have value. It would be a place of intrinsic value, a reminder of what had been and what could be again someday. People could draw inspiration from it when they needed inspiration. It would be waiting, inattentive to human needs yet serving them just the same.

Convincing people to value these woods was not easy, Saul knew. It took time and experience to appreciate these forests for what they were. Most people would see these magnificent trees simply as wood. Even longtime hikers, when tired or distracted, could just want to get back to their cars. Weyerhaeuser owned the land through which they were walking, even though it was in the national forest, and pressures were growing to cut the trees. The association had proposed that the valley be placed in a national monument so that the old growth would be saved. But to do that the association would have to give up ground elsewhere. There seemed no way to balance the needs of one against the other without giving up something irreplaceable.

The group hiked in clumps up the Green River valley, some talking, some walking silently. Partway up the trail they passed an old elk hunter's shack surrounded by fuel cans, dirty dishes, silverware, liquor bottles, beer cans, and plastic containers. Saul and McRae had complained to the Forest Service about the shack, but the agency was reluctant to do anything about it. The hunters who used the shack would object if the Forest Service removed it.

The hikers continued through the woods. They were intending to walk to the falls near the Minnie Lee mine—one of several old mines along the trail that had never panned out—and then return to their cars.

But one member of the group, Harry Deery, a retiree who had climbed Mount St. Helens twenty times and knew the area as well as he knew his own backyard, was deeply uncomfortable. Mount St. Helens is not visible from any part of the Green River. The ridge south of the valley rises steeply from the river and blocks any views in that direction. But Deery could sense the unseen volcano behind the looming green precipice. He knew that Saul and McRae had gotten assurances from the Forest Service that they were hiking in a safe place. But how could anyone know for sure that this valley was safe? He'd read an article in the paper saying that everything within fifteen miles of the volcano could be incinerated in an eruption, and they were well within that range. He'd seen paintings of historic eruptions—the people of Pompeii bombarded by falling lava from Vesuvius, the explosion of Krakatoa, Paul Kane's painting of Mount St. Helens erupting in 1847. With each step they were traveling farther away from their cars, farther away from safety, farther away from their only means of escape.

Eventually Deery couldn't stand it anymore. He told his wife that he was going back. She laughed at him—she wasn't about to give up this hike after driving all this way. But Deery couldn't continue. He turned around, walked by himself back to the cars, and sat waiting until the others returned.

UNDER THE VOLCANO

ON THURSDAY, MAY 15, COWLITZ COUNTY SHERIFF LES NELSON tried one last time to talk Harry Truman into leaving the Mount St. Helens lodge. He drove up the Spirit Lake Highway and parked outside the lodge with his motor running and the car pointed down the Toutle River valley. That Thursday was cloudy and cool, so Nelson didn't have to look up at the ash-draped bulge hanging off the north face of the mountain. But he could feel its presence behind the clouds, and it made the air heavy and oppressive. He knew that the weather forecast called for the weekend to be clear. Good weather meant that a lot more people would be around the mountain after a long wet spring, which would mean a lot more work for him and his deputies.

Inside, Truman was triumphant. A few days earlier he'd gotten a letter from Dixy Lee Ray. If he was ever thinking about leaving before, he certainly never would now. The letter read:

> Your independence and straightforwardness is a fine example for all of us, particularly for senior citizens.
>
> When everyone else involved in the Mount St. Helens eruption appeared to be overcome by all the excitement, you stuck to what you knew and what common experience and sense told you.

> We could use a lot more of that kind of thinking, particularly in politics.
>
> I get a fair amount of criticism for calling things the way I see them. I'm glad someone like yourself got credit for the same approach.

Why didn't the governor just come down and rip the badge off Nelson's chest? She was the one who was supposed to be in charge of public safety. If the governor wouldn't protect these people, then no one could. Truman had been hinting that he was ready to give up his vigil. He had told friends that he was tired of the earthquakes and couldn't sleep. To one, he had said, "The only reason I'm here is because they let me stay." A few days earlier, the sheriffs had scrambled to send a helicopter to the lodge when a Seattle news crew told them that they'd found Truman "in a broken-down and emotional state" and that he wanted to leave. But by the time the helicopter arrived, Truman had changed his mind. It would be hard for Truman to quit now, Nelson knew. He had painted himself into a corner with his bluster. People were expecting him to hold out against everyone telling him to go.

Nelson pleaded with Truman to leave but quickly realized it was futile. He said goodbye, got in his car, and drove as fast as he could back down the Spirit Lake Highway.

• • •

The next day, the concerns of the loggers working around Mount St. Helens came to a head. The geologists were saying that a landslide was inevitable, but the loggers were still working right next to the mountain. Even if a landslide affected just the valleys, those were the roads they needed to escape an eruption. One logger had already walked off the job on Thursday. More were threatening to do so the following week.

Friday morning, a safety representative for the International Woodworkers of America local named Joel Hembree drove up the south fork of the Toutle to talk with a group of about fifty disgruntled loggers who were working within five miles of the mountain's base. Several weeks earlier the company had promised that it would develop evacuation plans for each logging district. But in many cases, Hembree found, either the plans had not been developed or they had not been communicated to the crews.

Still, Hembree urged the loggers not to walk off the job. "We got the best experts in the world," he told them. "Supposedly you're going to get two hours' notice, but all I can tell you guys is if it blows, it blows. Who's to say it won't happen tomorrow, or ten years down the line?"

• • •

Also on Friday, John Killian called LeRoy Baine, who had been the best man at his wedding, to see if he and his wife wanted to come fishing with Christy and him over the weekend. The state had kept many of the lakes and rivers around Mount St. Helens closed to see what would happen with the volcano. But now that the eruptions had quieted down, it had opened them all, except for the lakes in the red zone. The fishing was bound to be great up there, John told LeRoy. No one had been up there all spring. The fish were just waiting for them.

But LeRoy's wife, Elna, had a cold; the Baines couldn't come. It was just as well. John and Christy were trying to start a family. If they were alone, they would have more time together.

Saturday morning, John and Christy stopped by the house where John's sister and her husband lived to pick up some fishing gear. John had always been close with his three sisters. He was their only brother. He had a responsibility to look out for them.

Charlene asked where he and Christy were going.

"They've opened Fawn Lake," John replied. "We're going there."

"Isn't it a bit close?"

"It's okay, Char."

Only in retrospect did Charlene wonder why she hadn't objected more strongly to John's plans. But there was no reason to be worried. The Weyerhaeuser Company had told their father that the loggers were safe working near the mountain, and Ralph had passed that information on to his crew and to John. The worst that was expected was flooding in the valleys and maybe some ashfall. Fawn Lake was nine miles and three ridgelines away from the volcano. They would be fine.

• • •

By this point, it had become obvious to local law enforcement personnel that the red and blue zones Dixy Lee Ray had established at the end of April were inadequate. The woods on the western, northwestern, and northern sides of the mountain were full of people who were working, camping, fishing, or trying to get a good look at the volcano. Because they were not inside the red or blue zones, police officers could not cite or fine them and had no authority to keep them out of the area. Weyerhaeuser had always kept most of the gates to its logging roads open, partly so loggers could get in and out of the woods quickly and partly because the company had a policy of keeping its lands open to the public. It was a generous act on the company's part, but now it was causing great headaches for the sheriffs and their deputies.

Les Nelson and the other sheriffs had been working with the state, the Forest Service, and Weyerhaeuser on an expansion of the restricted zone, and by May 15 all the parties had agreed on a proposal. It would extend the blue zone eleven and a half miles to the west and seven and a half miles to the north. It also would move the roadblock farther down the Spirit Lake Highway, to right above Weyerhaeuser's Camp Baker. The new proposal would actually make it easier for loggers, reporters, and property owners to get into both the blue and red zones. The media would be allowed into the red zone on helicopters so long as they did not land above the tree line. Loggers could get permits to log in the red zone, whereas before they had been allowed only in the blue zone. The main effect of the new zones would be to get the public out of the areas west and north of the volcano. The sheriffs would have legal authority to cite people who entered the area. And Weyerhaeuser would get the cars and campers off its logging roads and out of the way of its trucks.

That Thursday, Nelson sent a memo to the head of Washington State's Department of Emergency Services outlining the proposal. Extending the blue zone and moving the roadblock would place "the public farther away from the mountain," Nelson wrote, and would bring "the Governor's executive order into realistic alignment with existing geological conditions." Other sheriffs and public officials working in the area sent similar letters, citing the Geological Survey's warnings that the bulge eventually would give way.

State officials knew that the local sheriffs were right. But they delayed for a day while typing up the order and making sure that Weyerhaeuser had bought into the plan. The sheriffs were ready on Friday to move the roadblock and post the new blue zone before a weekend of warm and sunny weather. But the order never came. Not until Saturday morning did officials hand-deliver the order to the governor's office for her signature. By then, Ray was attending a Rhododendron Day parade in Port Townsend on the Olympic Peninsula. The new blue zone would have to wait until Monday.

• • •

Also by the third week of May, the owners of the eighty or so cabins a mile and a half west of Spirit Lake were in open revolt. Their cabins were inside the red zone and strictly off limits. They couldn't get to their property to remove personal items, do maintenance, or feed the pets they'd left behind. On Friday, May 16, a group of owners vowed that they would form a caravan the next day and drive to the roadblock on the Spirit Lake Highway to protest their exclusion. Rumors circulated that some planned to arm themselves and run the roadblock. The Washington State Patrol assigned eight extra cars to keep order.

When the owners assembled in the Toutle High School parking lot to begin the caravan Saturday morning—each wearing sky-blue sweatshirts bearing a drawing of Mount St. Helens and the words "I own a piece of the rock"—they learned that the governor would allow them to go to their cabins if they filled out a waiver and left the area by six that evening. Bending over the hood of a state patrol car, thirty-five people completed a form that absolved state and county agencies of responsibility for anything that might happen to them.

With two police cars in front and one behind, the caravan made its way up the Spirit Lake Highway to the cabins, accompanied by a procession of reporters. A state patrol plane flew overhead to keep an eye on the mountain.

The cabins were largely as their owners had left them. A thin layer of ash covered the ground. As predicted, skies had cleared Friday afternoon, and that Saturday was sunny and warm. Across the Toutle the cabin owners could see the ash-shrouded mountain, simultaneously grand and ominous in the afternoon sunshine. They loaded leather chairs, photos, fishing equipment, and other possessions into their pickups. One owner left a ten-pound bag of cat food open for her cat.

That afternoon, Rob Smith and his girlfriend Kathy Paulsen left the lodge the Smith family owned near the cabins and drove the rest of the way up the Spirit Lake Highway to see their friend Harry Truman. Truman was watering his lawn when they arrived, getting the lodge ready for summer tourists. Smith helped Truman sharpen his saws so he could cut more firewood. A police sergeant who was also visiting Truman gave him a bundle of mail, including letters from schoolchildren who had read about his refusal to leave the mountain. Truman's eyes teared up when he saw the letters. Some of his visitors that afternoon thought that he was just being sentimental; others, that the strain of holding out at Spirit Lake was getting to him.

As Smith and Paulsen prepared to leave, Truman followed them to their truck. Truman leaned in the window and said that he would see them the next day—he was planning to come to Castle Rock to buy primroses to plant in the garden. Smith and Truman both got teary when it came time to leave. "Oh, c'mon," said Truman, "let's keep a stiff upper lip."

By six p.m., everyone in the caravan had packed up their belongings. The state troopers made sure everyone was in their cars, escorted them back down the highway, and locked the gate behind. The governor had said that another caravan could travel to the cabins Sunday morning. In the small towns west of Mount St. Helens and up and down Interstate 5, property owners were making plans to retrieve their remaining possessions the next day.

• • •

That Saturday morning, the geologists in Vancouver had their usual staff meeting. Now that the volcano had calmed down, many were away that weekend. Crandell was back home in Denver. Mullineaux was in

California attending his daughter's college graduation. But the geologists and seismologists were still monitoring the volcano carefully. The bulge couldn't keep expanding forever.

At the staff meeting the geologists discussed a problem. For the past two weeks, a graduate student whom Dave Johnston had hired named Harry Glicken had been staying in the small white trailer parked at Coldwater II. His job was to monitor the bulge and issue a warning if he saw any sign of an avalanche or eruption. But Glicken had to leave for California that evening to talk about the graduate work he was starting in the fall, and the Geological Survey needed someone else at Coldwater II to keep an eye on the volcano. A geologist named Don Swanson, who had been working at the Hawaiian Volcano Observatory on the Big Island before Mount St. Helens became active, agreed to go. But after the meeting, Swanson sought out Johnston in a hallway. Could Johnston fill in for him that evening? Swanson had a graduate student from Germany visiting, and he wanted to see the student off on Sunday morning. Swanson would replace Johnston as soon as the student was gone.

Johnston was much more worried than either Swanson or Glicken about spending time near the volcano. After his close escape from Mount Augustine, he knew that being near a volcano could be deadly. And he'd never liked the looks of Mount St. Helens. It was doing things that volcanoes weren't supposed to do, things that were difficult to understand. At this point, the geologists agreed that the bulge had to be the result of magma welling up inside the volcano and pushing on the mountain's north flank. But if that was so, why hadn't Johnston detected higher levels of sulfur emerging from vents on the volcano's sides? It was as if the magma were somehow bottled up and unable to escape to the surface. But did that mean that the mountain was about to erupt or that it was more stable than they'd thought?

Despite his misgivings, Johnston agreed to fill in for Swanson that night as long as Swanson would replace him on Sunday. Later that day he drove his tan government-issued Ford Pinto station wagon up the Spirit Lake Highway. Johnston had never been to Coldwater II, and he missed the turnoff to the observation post and continued up to Timberline. There he boarded the helicopter the geologists had been using for their research and flew high onto the side of the volcano. He jumped out and quickly measured the temperature of a steam vent. It was only 190 degrees Fahrenheit, which Johnston judged a "poor temperature" for a fumarole. But then rocks began falling all around him as an earthquake shook the mountain. Johnston dashed back to the helicopter and flew away.

Later that afternoon, two young scientists with the Geological Survey drove up to Coldwater II to visit with Johnston and Glicken, who had not yet left for California. Mindy Brugman and Carolyn Driedger were studying glaciers on Mount St. Helens and other western volcanoes. Brugman had become an expert with the device the geologists were using to measure the distance between Coldwater II and the bulge, and she wanted to see if Johnston was having any trouble with it. He wasn't, so Brugman and Driedger simply settled in to enjoy a beautiful Saturday afternoon. They had seen other cars and trucks on the ridgelines as they'd driven in, mostly people who had taken logging roads to get near to the volcano. But the volcano seemed quiet. Even the growth of the bulge had slowed slightly. The four geologists took turns photographing one another in the director's chair outside the white trailer.

As the sun dropped toward the western horizon, Brugman and Driedger asked if they could spend the night on the ridge. They had their camping gear in the back of Brugman's truck. They wouldn't be any trouble. But Johnston insisted they leave.

"Why?" asked Driedger. "Rocky's showed that nothing has ever happened up on this ridge."

"That doesn't mean it couldn't happen," Johnston said. "When that landslide comes down, it could come all the way across that valley and up over the top of this ridge."

"But the mountain's five miles away," Driedger objected.

"It could happen."

About seven that evening, Brugman and Driedger threw their stuff in the pickup and headed back down the logging road. Brugman remembers that she was surprised by how many animals kept jumping into the roadway.

Before he went to sleep, Johnston got on the radio with a geologist named Dan Miller in Vancouver. Miller told him that the armored personnel carrier was on a flatbed truck traveling down Interstate 5 and would arrive the next day.

"Are you serious?" asked Johnston, who had not heard about the plans for the personnel carrier.

"I'm serious," Miller replied.

"Are they going to provide ammunition?"

"That's negotiable at this point," said Miller.

• • •

By that evening, Harry Truman was the only person left on the shores of Spirit Lake. But in a ten-mile by thirteen-mile rectangle north of the volcano, more than twenty other people were also settling down for the night. If the blue zone had been extended to the north and west, as the local sheriffs were advocating, most would not have been there. But with the extension order sitting unsigned on Dixy Lee Ray's desk, none of them was doing anything wrong.

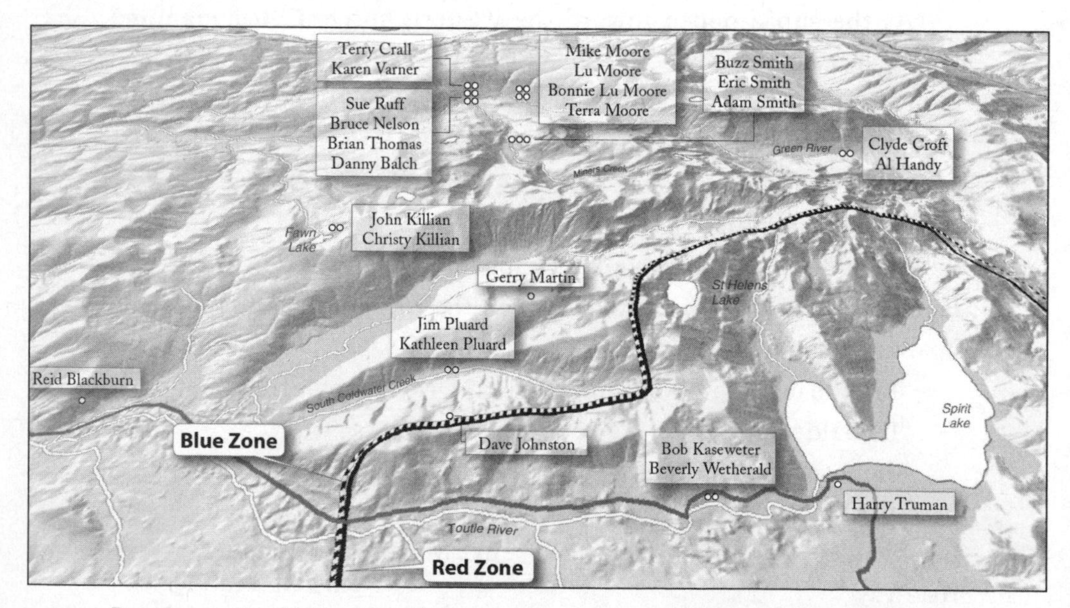

People north of Mount St. Helens on the evening of Saturday, May 17, 1980

In the cabins west of Spirit Lake, the state patrol had ushered everyone back down the Spirit Lake Highway except for Bob Kaseweter, a thirty-nine-year-old geochemist, and his girlfriend, Beverly Wetherald. They had a permit allowing them to stay in their cabin, even though it was in the red zone. They both worked for the Portland General Electric utility company, and Kaseweter had convinced the authorities to let him conduct research on the mountain from his cabin. He had rigged up a battery-run seismograph, with the graph paper disappearing through a hole in the floor into the basement. He also was photographing the bulge from his deck, documenting the progressive widening and lengthening of the cracks in the ice.

Some people weren't happy about the couple being allowed to stay. Other cabin owners complained that Kaseweter's studies were an excuse to avoid the red-zone restrictions. "It's a serious study," Kaseweter retorted. "I'm not using it as an excuse. It's the chance of a lifetime."

Kaseweter and Wetherald had driven up the Spirit Lake Highway Friday afternoon in Kaseweter's orange Volkswagen but had gotten stopped by a state trooper at the roadblock. "I'm doing geology at my cabin," Kaseweter said. "This is my secretary."

The trooper got on the radio to check out his story. About then, Harry Truman arrived at the roadblock in his white pickup.

"Everyone says you know about the volcano," the trooper said to Truman while waiting to hear back on the radio.

"Hell, I've no idea what the damn thing will do," said Truman.

The radio came on telling the trooper that Kaseweter and Wetherald could go, but not in their car. "Okay to pedal up?" asked Kaseweter, lifting a bicycle from its rack on the Volkswagen. He could leave the car at the roadblock and pick it up when they left.

"How'm I going?" asked Wetherald.

"I'll give you a ride," said Truman. "Jump in."

• • •

Coldwater II was high on the ridgeline a mile northwest of Kaseweter's cabin. From there, Johnston and the other geologists looked across the broad, wooded valley of the Toutle River to the ash-covered bulge on

Mount St. Helens's northern flank. But they weren't the only people on nearby ridgelines that evening.

Washington State did not have the funds to pay for observers who could warn surrounding communities of an impending eruption, so it turned to a volunteer organization. The Radio Amateur Civil Emergency Service had helped the state previously with fires and floods. Now the organization had agreed to station ham radio operators with mobile rigs around the mountain to monitor its activities.

Earlier on Saturday, Gerry Martin, a sixty-four-year-old volunteer ham radio operator from Concrete, Washington, had driven his twenty-six-foot green-and-white Dodge Superior motorhome up a rough gravel road and onto the ridge just north of the Coldwater II post. A retired US Navy radioman, Martin parked in a clearcut a bit higher than Dave Johnston, so he could see, across the valley containing South Coldwater Creek, both the mountain and the trailer in which Harry Glicken had been sleeping the past two weeks. Other members of the radio network also were stationed around the mountain that Sunday morning, but Martin was the closest.

One other person was on a ridgeline north of the mountain that evening. At the Coldwater I observation post two miles west of Johnston, photographer Reid Blackburn was taking his last photographs of the day. Blackburn, a twenty-seven-year-old photographer for the Vancouver *Columbian* newspaper, was participating in a project arranged by Fred Stocker, a freelance photographer for *National Geographic* magazine. Blackburn's job was to trigger by remote control a series of cameras stationed around the mountain. The original plan called for six cameras, but only two turned out to be available—the others were lent out to cover the Kentucky Derby. One camera was set up with Blackburn at Coldwater I. The other was on the side of Mount Margaret north of Spirit Lake.

Blackburn was popular in the *Columbian* newsroom. He liked to play jokes on his officemates, like filling the photo editor's coffee cup with cigarette butts during an anti-smoking campaign. He had thick brown hair and a well-kept beard and wore a pair of square black glasses. The *Columbian* had been more cautious than other media outlets, which had been playing up the story through stunts like flying reporters and photographers to the rim of the crater. Blackburn's location was supposed to minimize the danger while giving him a clear view of the volcano.

Blackburn was perfectly suited for the job. He was a skilled photographer who thrived in a place of great natural beauty like southwestern Washington. He also was an outdoorsman who often camped and hiked around Mount St. Helens. He and his wife, who worked in the advertising department of the *Columbian*, had gone cross-country skiing there on March 23, three days after the first earthquake but before the area around the mountain was closed. He was a licensed ham radio operator, which was required to operate the radio-controlled cameras. And he was accustomed to dangerous assignments—a few months earlier, he had photographed a wrecked tanker car leaking ammonia gas, betting that the wind wouldn't change directions and coming back with dramatic photos of the tanker's engineer and brakeman being rescued.

On Saturday, Stocker and the driver he had hired for his *National Geographic* assignment, a man named Jim McWhirter, spent much of the day with Blackburn. But McWhirter was restless. "The whole day I'd just been on edge," he later recounted. "I don't know what the feeling was. Now I attribute it to the Holy Spirit telling me to watch out, even though I didn't really believe in God at the time. I just had to do something." Blackburn told them that he was going to stay; he had an obligation to trigger the cameras. Finally Stocker suggested that he and McWhirter drive to his girlfriend's restaurant near Olympia for a steak and lobster dinner. McWhirter quickly agreed, but he said that he wanted to be back by eight the next morning, since he had no desire to spend the entire next day with Stocker and his girlfriend. That night they ended up "partying pretty hard," McWhirter said, and didn't get to bed until three a.m.

On Thursday, Blackburn had talked by phone with his wife, Fay, about the project. Placing a phone call from Coldwater I was complicated. Blackburn had to have an operator patch his ham radio through to the phone network, and regulations limited phone calls to just three minutes. In their brief time on the phone, Blackburn and Fay talked about the project. The newspaper was about to shut it down. The remote-controlled cameras were producing essentially identical photographs, and nothing seemed to be happening with the volcano. But he would stick it out over the weekend, Blackburn said. He enjoyed camping at Coldwater I. He could stay warm near the fire barrel the geologists had left behind. He listened to '70s music on a cassette tape he'd made.

After Stocker and McWhirter left on Saturday, Blackburn placed another call to Fay. She heard the phone ringing just as she was leaving the house, but by the time she answered, the caller was gone.

• • •

After stopping by Charlene's house, John and Christy Killian drove up the Green River and through a tangle of logging roads that John knew as well as he knew the streets of Vader. Fawn Lake rests in a classic alpine cirque carved by an Ice Age glacier into the north side of the ridge, where the Killians were camping. It has the shape of an amphitheater, with a high wood-covered slope to the south. On the north end of the lake, near its outlet, the land drops off into the valley that cradles Shultz Creek. Weyerhaeuser had heavily logged the area around the lake, but the steep slopes of the cirque and the area around the lake's outlet to Shultz Creek, where the Killians camped, was still covered with old-growth trees.

The Killian family had been coming to Fawn Lake for decades. It was one of their favorite places to camp and fish in all of southwestern Washington. Before they went to bed that evening, John and Christy undoubtedly enjoyed the firm, fresh flesh of eastern brook trout.

Northeast of Fawn Lake, Edward "Buzz" Smith and his two sons—Eric, age ten, and Adam, age seven—also were getting ready to crawl into their tent. Smith was a Weyerhaeuser logger from Toutle who knew the backcountry well. He and his sons had driven up Road 2500 to where it crosses the Green River and heads into the high country. There they parked in the flat area where Susan Saul and her associates had parked the week before. They hiked up the Green River a mile or so. Then they crossed the Green on a log bridge and followed Miners Creek to the base of a large bluff. They set up their tent just to the west of Black Mountain, one of the highest peaks in the country north of Mount St. Helens.

• • •

Back on the Green River, three separate groups had set up camps that Saturday night.

Farthest upriver, Clyde Croft, thirty-seven, and Al Handy, thirty-four, had laid out their sleeping bags beneath a blue plastic tarp. They were friends who worked together in a warehouse of the West Coast Grocery Company in Tacoma. They were very different men. Handy was quiet and introverted; Cross, outgoing and rambunctious. But they enjoyed each other's company, both at work and when they got together after work.

For nearly a year they had been planning a horseback-riding trip north of Mount St. Helens. Now that the first nice weekend of the year had arrived, they had decided to go. On Friday, Croft had borrowed a pickup truck and horse trailer to cart two of his horses up the mountain. After finishing work about midnight on Friday, they headed out. They drove east on Highway 12 to a town called Randle halfway between Mount Rainier and Mount St. Helens. There they turned south and followed a rough set of logging roads to Ryan Lake, a small, clear lake about twelve miles north-northeast of the volcano. To the south of the lake, the Green River descends from its origins in the high country north of Mount St. Helens. To the west, the Green heads toward the Toutle, the Cowlitz, the Columbia, and the sea.

Saturday morning they saddled their two horses, a thirteen-year-old Bay named Cochise and an eleven-year-old Appaloosa named April. They spent the day riding the steep and narrow trail to Deadman's Lake, 1,000 feet above Ryan Lake—a gorgeous alpine pool with granular pumice beaches and a dark blue surface. By that evening, they were back in the Green River valley. There they camped near the Polar Star Mine, an abandoned shaft on a Forest Service road about two miles west of Ryan Lake. Croft had plans to meet his brother on a Tacoma golf course the next morning.

Farthest downriver, six friends from the Longview area were eating roasted elk and baked potatoes cooked in the coals of their campfire. Terry Crall and Karen Varner were both twenty-one, tall, blond, and good-looking—people often mistook them for twins. They had met at a party three years before when Terry had asked Karen to dance. Since then they had been inseparable. They had been living together and were talking about getting married. But their lives were in flux. A month before Terry had been laid off from his job at a Weyerhaeuser lumber mill. Karen was working as a receptionist at a doctor's office but wanted to return to community college and become a nurse. Still, they didn't take their jobs too seriously. They spent as much time outdoors as possible. They had hundreds of friends.

One of those friends was Sue Ruff, who was there camping with her boyfriend, Bruce Nelson. They too were a handsome couple. Bruce had shaggy brown hair; Sue had braids that fell past her shoulders. Sue was the common denominator among the six people in their party. She knew Terry and Karen, and she also had asked her friend Brian Thomas to come along on the trip. The sixth member of the group was Danny Balch, who had been planning to go to the beach that weekend. But Brian wanted a friend to come on the trip so he wouldn't be alone with the two couples, and Danny had agreed to come to the Green River instead.

The six of them had driven up Road 2500 and parked near Buzz Smith's pickup. They hiked just a few hundred yards up the trail to a set of tent sites near the dilapidated cabin that Susan Saul and her group had passed the week before. There they settled in for a long fun evening.

The third party on the Green River was camped partway between the Longview group and Croft and Handy. Earlier that week, the Moores—Mike and Lu—had decided to take their daughter, Bonnie Lu, age four, on her first backpacking trip; three-month-old Terra could come along in Lu's baby backpack. Mike, a geology major in college and an ardent backpacker and camper, had chosen the hike carefully. He wanted a well-groomed trail, an easy grade, and a beautiful setting. On all three counts, the Green River was perfect. Mike and Lu had been there often, and they wanted to show it to Bonnie Lu. And it seemed perfectly safe. The valley was thirteen miles from the volcano, with a 6,000-foot jumble of mountains separating them from it. Mike had been paying close attention to the volcano's activity over the previous two months. He didn't expect it to be a problem.

Saturday afternoon the Moores drove their car up Road 2500 and found three other cars in the parking lot by the river—Buzz Smith's Chevy pickup, Bruce Nelson's Blazer, and Danny Balch's Ford. They laced up their boots, shouldered their packs, and headed up the trail. They were planning to set up their tents where Sue Ruff and her friends had camped, but those campsites were already taken, so they kept walking into the middle of the old-growth trees that Weyerhaeuser had not harvested. There they set up camp amidst one of the most beautiful forests on earth.

• • •

As the sky darkened Saturday evening and a thin sliver of moon followed the sun toward the Pacific Ocean, a hush fell on Mount St. Helens—the hush of a dense woods under a clear, cold, moonless sky. It had been a gorgeous day, the kind of day that buoys the spirits of Northwesterners beaten down by a cold, rainy spring, and the forecast for Sunday promised equally beautiful weather.

PART 4
THE ERUPTION

Mount St. Helens at 8:33 a.m. on Sunday, May 18, 1980

KEITH AND DOROTHY STOFFEL

KEITH STOFFEL COULDN'T BELIEVE HIS GOOD LUCK. THE DEPARTment of Natural Resources geologist and his wife, Dorothy, had driven from Spokane to Yakima the previous day so he could give a lecture about Mount St. Helens at a gem and mineral show, and the weather forecast had promised that Sunday would be clear. He had called the airport before the lecture and reserved a Cessna, and Sunday morning he, Dorothy, and the pilot, a man named Bruce Judson, were flying through a cloudless sky toward the glistening white peak on the horizon. Almost everyone else in the department had already flown over the mountain, but Keith, who was twenty-seven, and Dorothy, age thirty, a consultant for a private geology firm, had not yet seen it.

Dorothy had never been on a small plane before, and she had been terrified that morning before they took off. "Do you like to fly?" she asked Judson as they were walking across the tarmac. "It's just like any other job," said the twenty-three-year-old, trying to reassure her. "It gets boring after a while."

They were on their fourth pass over the north rim of the crater, flying west to east, when Keith noticed something moving. "Look," he said, "the crater." Judson tipped the Cessna's right wing so they could get a better view. Some of the snow on the south-facing side of the crater had started to move. Then, as they looked out the plane's windows, an incredible thing happened. A gigantic east-west crack appeared across the top of the mountain, splitting the volcano in two. The ground on the northern half of the crack began to ripple and churn, like a pan of milk just beginning to boil. Suddenly, without a sound, the northern portion of the mountain began to slide downward, toward the north fork of the Toutle River and Spirit Lake. The landslide included the bulge but was much larger. The whole northern portion of the mountain was collapsing. The Stoffels were seeing something that no other geologist had ever seen.

A few seconds later, an angry gray cloud emerged from the middle of the landslide, and a similar, darker cloud leapt from near the top of the mountain. They were strange clouds, gnarled and bulbous; they looked more biological than geophysical. The two clouds rapidly expanded and coalesced, growing so large that they covered the ongoing landslide. "Let's get out of here," shouted Keith as the roiling cloud reached toward their plane. Judson put the Cessna into a steep dive to pick up speed. He redlined the airspeed indicator; any faster and he would rip off a wing. Keith noticed that the cloud was extending to the north and east. "Turn right!" he yelled.

Now flying south, they began to outrun the cloud. From the front seat, Dorothy turned and looked back at the volcano. A column of dark ash, like a solid object, rose from the mountain higher than she ever thought possible. It was shot through with lightning—pink and purple and yellow. Far overhead, the ash had spread into an anvil shape and was beginning to blow east. They heard no sounds other than the whining of the Cessna's engines. It was as if the volcano were erupting silently.

HARRY TRUMAN, BOB KASEWETER, AND BEVERLY WETHERALD

THE THREE PEOPLE NEAREST MOUNT ST. HELENS—HARRY TRUMAN in his lodge on Spirit Lake, and Bob Kaseweter and Beverly Wetherald in their cabin a mile downstream—were close enough to hear a rumbling from the mountain as its north flank began falling toward them. But before the avalanche reached them, the expanding cloud that the Stoffels had seen from the plane overtook the falling earth and raced ahead. The cloud sped down the flank of the volcano, ripping the forest from the mountainside as it passed. When the cloud hit the resorts and camps around Spirit Lake and the cabins on the Toutle River, it blew the structures to bits.

A few seconds later, the avalanche reached Spirit Lake and the Toutle River. It buried the lodge and cabins under hundreds of feet of steaming stone, earth, ice, and mud. Harry Truman, Bob Kaseweter, and Beverly Wetherald were dead before they could have known what was happening.

DAVE JOHNSTON

ON THE RIDGE OVERLOOKING THE TOUTLE RIVER, DAVE JOHNSTON had been awake almost since the sun rose at 5:36 that morning. He had already measured the distance to the bulge three times, registering first a slight expansion and then a slight contraction.

"What are things like up there?" Johnston's colleague Bob Christiansen asked over the radio.

"It's very nice," said Johnston. "You can see the mountain clearly. There are no clouds." They talked for a moment about Johnston's gas-monitoring data. "Okay, that's all I have to say," Johnston said. "Coldwater, clear."

Johnston was watching the mountain when the landslide began. He must have watched it for at least a few seconds. The landslide consisted of three blocks that slid away from the mountain. When the first block dropped toward the Toutle River and Spirit Lake, it exposed the top of a pocket of magma that was partly embedded in the landslide's second block. This intruded magma had been pushing outward under the mountain's north flank, creating the bulge. But, as Johnston had suspected, the overlying rocks had sealed the magma up tight, heightening the pressure inside the mountain. When the landslide released that pressure, the water in the magma flashed into steam. That's what created the two rapidly expanding clouds that the Stoffels had seen from their plane. The magma pocket was like a steam boiler. For almost three months the pressure inside that boiler had been building. Now the boiler had failed.

Technically, the explosion that swept out of Mount St. Helens that morning is known as a pyroclastic density current. Most geologists refer to it simply as "the blast," though some prefer the term "surge," contending that it was not really an explosion. The blast cloud accelerated as it spread, drawing heat energy from the fragmented magma it contained. Inside the cloud were ash, pumice, lava blocks, snow, ice from the overlying glaciers, tree fragments, soil swept from the ground, and boulders as big as cars. It expanded at speeds of hundreds of miles per hour, but in a particular way. As Dave Johnston and Barry Voight had feared, Mount St. Helens did not explode straight up. It exploded to the side, in the direction of the bulge. The avalanche created an amphitheater-shaped gouge in the mountain, and this gouge channeled the blast to the northwest, north, and northeast. It was as if the blast had emerged from the muzzle of a cannon pointed directly at Coldwater II.

When Johnston saw the north flank of the mountain give way, he flipped on the radio. "Vancouver, Vancouver! This is it!" he said. Most people have interpreted this statement to mean that Johnston realized that the big eruption, the one he had feared all along, had finally occurred. But Johnston was a scientist with an assignment, and his voice was excited, not fearful. He was there to monitor the volcano for precursors that could signal the beginning of a landslide down the Toutle River valley. He was probably trying to tell his fellow geologists in Vancouver that the landslide they had expected had finally begun.

As Johnston watched the volcano, the blast cloud quickly obscured the ongoing avalanche. The front of the cloud was magnificent. It was like an immense oncoming waterfall with great blocks of earth and ice cascading from far overhead. The base of the blast cloud reached the ridge on which he was standing and started up the side. Johnston could have tried to take shelter. He might have run inside the trailer, thinking he could survive. But by the time the blast reached the ridge, he must have known he wouldn't live. It is certainly possible that he stood and watched the oncoming holocaust.

When the blast front hit Coldwater II, it flung the trailer, the Ford Pinto, and Johnston across the top of the ridge as easily as someone would brush away a fly. Johnston undoubtedly lost consciousness and died almost immediately. Everything that was on the top of the ridge flew into the adjacent valley.

Johnston had been right when he was talking with Carolyn Driedger the previous evening. When the debris avalanche generated by the landslide reached the thousand-foot ridge on which Johnston had been standing—a few seconds after the ridgetop was raked clean by the blast cloud—the flowing debris swept up and over the ridge and into the valley beyond. Johnston, his car, and the trailer were buried and have never been found.

On the wall of his childhood bedroom in his parents' house in Illinois, Johnston had tacked a quotation from Teddy Roosevelt that he had carefully copied as a teenager. It read:

> It is not the critic who counts; not the man who points out how the strong man stumbled or where the doer of deeds could have done them better. The credit belongs to the man who is actually in the arena, whose face is marred by dust and sweat and blood; who strives valiantly; who errs and comes short again and again; who knows the great enthusiasms, the great devotions; who spends himself in a worthy cause; who, at the best, knows in the end the triumph of high achievement, and who at the worst, if he fails, at least fails while daring greatly, so that his place shall never be with those timid souls who know neither victory nor defeat.

JIM AND KATHLEEN PLUARD

NOT EVERYONE KILLED NEAR MOUNT ST. HELENS HAD CAMPED there the previous night. Jim Pluard—the foreman of the tree cutters who had objected back in April about logging so close to the mountain—liked to check on his crew's logging equipment over the weekend. Weyerhaeuser had a lot of money invested in its equipment, and Pluard wanted to make sure it hadn't been vandalized and would be ready to go Monday morning. He owed a lot to Weyerhaeuser. His father had been a logger, and now two of his sons were in the business. Weyerhaeuser paid a good wage and treated its employees well. The least he could do was keep an eye on things over the weekend.

Sunday morning, Pluard and his wife Kathleen tacked a note to the front door of their home in Toledo. It read: 7:30. GONE TO THE MOUNTAIN. BACK IN TWO HOURS. They got in Pluard's company pickup and headed up the Spirit Lake Highway. If they left Toledo at seven thirty, they would have gotten to the logging site on the back of the ridge where Johnston was stationed about an hour later.

Like Johnston and the trailer, the Pluards and their pickup were never seen again. Dorothy Stoffel later said that she saw a red pickup driving toward Timberline right before the volcano erupted. But the company pickup Pluard was driving was yellow. Perhaps the color looked different from a height. Otherwise, the occupants of the red pickup remain a mystery.

GERRY MARTIN

ON THE RIDGELINE BEHIND JOHNSTON, HAM RADIO OPERATOR GERRY Martin also had been up for hours. Three cats traveled with him in his motorhome, and he walked them on leashes so they wouldn't run away. He likely had already taken them out that morning. By 8:30 he was in the driver's seat of his motorhome monitoring the mountain.

At 8:32, he radioed to the network, "Oh oh, I just felt an earthquake, a good one, shaking, uh, here's a—" At this point his transmission was cut off by a device that limits the length of transmissions.

Another ham radio operator cut in to tell Martin that he should change his frequency for better reception.

Twenty seconds later, Martin resumed his broadcast: "Now we've got an eruption down here. Now we got a big slide coming off. The slide is coming off the west slope. Now we've got a whole great big eruption out of the crater. And we got another opening up on the west side. The whole west side, northwest side, is sliding down."

A few seconds later Martin continued: "The whole northwest section and north section blown up, trying to come up over the ridge towards me. I'm gonna back outta here." He spoke calmly, reflecting his years of navy training.

Martin was turning the key in the ignition, yet he remained on the radio. His words were interrupted by static from the lightning in the blast cloud. He watched the cloud envelop the trailer at Coldwater II. "Gentlemen, the camper and car that's sitting over to the south of me is covered. It's going to hit me too." At this point, the blast was traveling near its maximum speed, which meant that it took about twenty seconds to travel the two miles from Johnston to Martin.

Again, Martin's microphone switch opened and closed, but he said no more.

REID BLACKBURN

LIKE JOHNSTON AND MARTIN, PHOTOGRAPHER REID BLACKBURN arose with the sun that Sunday morning. He took photographs with the two cameras at 7:11, jotting in his notebook the number of shots taken and the time. Blackburn had always been a careful worker, and documentation was essential in his line of work.

The next line in his notebook is marked with the time 8:33. Blackburn was taking photographs of the ongoing eruption. He wrote quickly; his handwriting was shaky. He snapped his notebook shut, threw it in a container with his ham radio, and lowered and latched the container's lid. Then he jumped in the front seat of his car.

The blast cloud reached him before he could put the key in the ignition. The car windows facing the volcano blew out. The Volvo quickly filled with burning hot ash. Blackburn tried to breathe, but the blast cloud contained little oxygen. His nose, mouth, and lungs filled with ash. Ash from Mount St. Helens tastes like chalk dust mixed with metal; it smells like a dry field stirred by the wind on a hot day. Those must have been among Blackburn's last sensations as he died.

JOHN AND CHRISTY KILLIAN

BECAUSE THE BLAST CLOUD CONSISTED LARGELY OF CRUSHED PUMice and pulverized lava, it was initially much heavier than the surrounding air. As a result, it moved more like a fluid than a gas. It hugged the ground as it flowed away from the volcano. When it encountered a ridge, it swept up one side and down the other. Climbers on Mount Rainier fifty miles to the north said that the blast looked like a dark wave flowing over the countryside.

When the volcano exploded Sunday morning, John Killian was almost certainly on Fawn Lake fishing. Before leaving Vader the previous day, he had thrown a rubber raft into the back of his pickup, and scraps of the raft were later found far down the valley below the lake. Christy was at the campsite with their pet poodle.

From where Christy sat at the outlet of Fawn Lake—nine miles away from the summit of Mount St. Helens—the blast cloud rose above the ridge across the lake like an immense oncoming storm. Then it spilled over the edge of the hill and swept across the water. Instantly, every tree around the lake snapped. Christy was caught in a maelstrom of wood, stone, water, and hot ash. She was torn to pieces. When her left arm was found several months later, it was identified by the wedding band on her finger.

John's experience was different. Sitting on the edge of the raft, he would have noticed the surface of the water ripple as the air pressure suddenly changed. Then the oncoming blast picked up the raft on which he was sitting as easily as if it was a leaf on a pond. For the last instant of his life, John Killian was flying, flying through the liquid air.

CLYDE CROFT AND AL HANDY

AT 8:30 ON SUNDAY MORNING, CLYDE CROFT AND AL HANDY, WHO had camped near the Polar Star Mine, were preparing their horses for the journey back to Ryan Lake. It was a quiet morning, broken only by the murmuring of the nearby Green River. Suddenly the air pressure began to change, making their ears pop. When they looked to the south, they saw a great black wall of ash rising above the ridgeline. Then the wall of ash crested the ridge and began to descend toward their campground.

The Green River valley marked the northern edge of the blast zone. By the time it reached the Green River, the blast cloud had deposited so much coarse ash over the countryside, and it had expanded so greatly because of the heat it contained, that it became less dense than the surrounding air. It began to rise, like smoke from a fire. A countervailing wind sprang up from the north, pushing the blast cloud higher. The wall of ash moved laterally back and forth across the Green River, first in one direction and then the other. Some parts of the Green River valley were engulfed. Others were virtually untouched.

Croft and Handy had camped in a part of the valley that was engulfed. Handy raced up the hillside north of the river, hoping to take cover in the mine. He was about twenty yards away from the entrance when the ash enveloped him. The hot ash filled his throat and seared his lungs. Within seconds, Handy and the two horses died of ash asphyxiation.

Croft made a different decision. He grabbed a sleeping bag and dove into the Green River. Above him, hot ash fell onto the sleeping bag. Below him, the frigid water of the Green River swirled around him. For more than an hour he remained partially submerged in the water with the sleeping bag over his head. It was completely dark. There was nothing to do but wait.

Finally the darkness lifted enough for him to rise from the river. He began walking up the trail to Ryan Lake, holding the sleeping bag over his head. Croft had been a combat veteran of the Vietnam War and had survived a hard childhood on a poor Texas farm. He was strong and determined. Burns covered his face, arms, and chest, but he kept going. Nothing was more important than continuing to move.

His breath became labored as he inhaled more and more ash. Finally he reached the truck at Ryan Lake. It was covered by fallen trees, unusable. He pulled out a case of warm Olympia from the front seat and quickly downed two cans of beer. But it hurt to swallow. The hot ash had burned his throat and lungs.

Croft set off again on the road toward Randle, twenty-five miles to the north. He didn't have to get all the way there. Someone would rescue him if he could get away from the ash. Two miles north of Ryan Lake, he began to grow dizzy. He weaved from side to side as he walked. He came upon a large uprooted tree. He tried to climb over the tree, but it was too large and he was too weak. He dropped to his knees and began digging through the hot ash, forming a tunnel so he could crawl underneath. He made it to the other side, got back on his feet, and kept walking.

Finally he made it past the blown-down trees. Tractors and other machinery were parked on the side of the road. Croft got into one, and turned the key in the ignition, but the tractor would not start. Because of the heavy ashfall, all of the vehicles were unusable. He gave up and continued down the road.

For three more miles he walked. If he kept going, he would make it. But he had to rest. He had walked eight miles, badly burned, with his lungs half full of ash. He moved to the side of the road, laid his sleeping bag in the ditch, and wriggled inside. He rested his head on his arm. Slowly his breathing quieted, and then stopped.

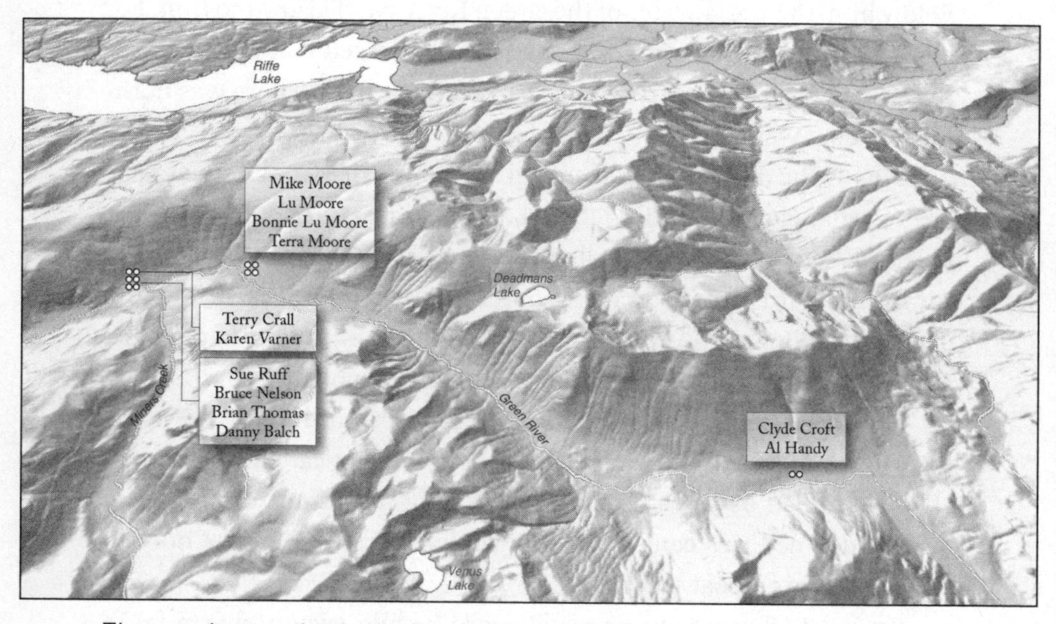

The people camping in the Green River valley Sunday morning

TERRY CRALL, KAREN VARNER, BRUCE NELSON, SUE RUFF, BRIAN THOMAS, AND DANNY BALCH

ON SUNDAY MORNING, TERRY CRALL LEFT HIS GIRLFRIEND, KAREN Varner, in their tent and went fishing for steelhead in the nearby Green River. In the tent with Karen was Tye, an Australian shepherd, and Tye's three puppies. Nearby, Bruce Nelson and Sue Ruff huddled around the fire trying to get warm. The Green River is nestled so deeply amid the ridges north of Mount St. Helens that it takes the sun a long time to reach the valley floor. They had stuck some leftover marshmallows on sticks and were waiting for a pot of water to boil to make coffee. About 70 yards down the trail toward Road 2500, Brian Thomas and Danny Balch were still sleeping in a tent after staying up late drinking beer around the campfire.

About eight thirty, Terry came bursting into the campsite to tell Sue and Bruce about a huge fish that had broke his line. As Sue rose to get a pack of Camel Lights from the tent, she noticed a small plume of smoke rising above the southwestern tree line. "There must be a fire somewhere," she said. She considered taking a picture of the strange plume of smoke, but as she stared at it, it began to change.

Quickly the plume filled the sky to the south and west, with odd-looking stalks extending upward and outward. Suddenly a gust of cold air blew through the campsite. Their campfire shot sideways, and Sue's braids blew straight out from her head. The cloud above them quickly turned from yellow to red to black. Within seconds, hurricane-force winds were blowing through the trees.

Terry ran to the tent where Karen and the dogs were sleeping and jumped inside. An instant later, a large tree fell directly on top of them.

Bruce wrapped his arms around Sue. They were standing between two large Douglas fir trees. All around them, trees were falling, creating great booming sounds like cannons being fired. The tree standing next to them fell, and the couple toppled into the hole where the roots had been. Above them, falling trees partly covered the hole. They were engulfed by darkness. They clutched each other as the air became warm and then burning hot, as though they were in a giant furnace. Bruce, who'd been a baker, estimated that the temperature reached more than 500 degrees Fahrenheit. All the hair on their arms was burned off, and they could feel the hair on their heads being singed.

"Are you okay?" Bruce whispered.

"Okay," Sue replied. Bruce could barely hear her, even though her face was inches from his own.

"My God, Sue," he said. "We're dead." Chalky grit filled his mouth.

"Bullshit," she replied, "we're not dead yet."

Nearby, Brian and Danny had both leapt out of their tent when Brian, sensing that something was wrong, had looked out the back of the tent and had seen the blast cloud approaching. Instantly the cloud was upon them. The trees around the campground all fell at once, as if yanked from the ground by an invisible hand. Brian rolled behind a log that had already fallen. In seconds he was buried by ash, branches, and debris. The blast pushed Danny to the ground. He was pelted by mud and ice, which melted as it hit him. Then he was hit by a wave of scorching heat. He flung out blindly, trying to find anything to grab on to. His hands grasped some burning logs, and he screamed in pain as the skin on his hands melted. Another terrible burn covered his left leg. He called for Brian but got no response. Thick clumps of dirt began to fall in the pitch-black darkness.

In the root pit farther up the trail, scalding ash was falling on Sue and Bruce. They had to dig the chalky gray grit from their mouths with their fingers. They pulled their shirts over their heads to protect themselves from inhaling too much ash. They felt themselves getting cold, sleepy, and nauseous, though whether because of gases in the blast cloud or shock they couldn't tell.

Fifteen minutes after the blast hit, the cloud shifted to the south and the sky lightened. Still, chunks of coarse particles, ice, and vegetation kept falling from above and hitting the fallen trees. Slowly, Bruce and Sue began to dig their way out of the root pit, gagging on the ash-filled air. They called out to their friends but heard no reply. Sue managed to take out her ash-coated contact lenses.

For an hour and a half they huddled behind a fallen tree. Visibility was low. The air smelled of brimstone, like a hot spring. There were no signs of life. "If we get out of here alive, you're going to marry me," Bruce said. Sue said that she would.

When the ash cloud had lightened, Danny had made his way to the river to soothe his burned hands and leg. Then he set out to search for Brian. He was struggling through a tangle of trees when he felt something grab his leg. It was Brian. He had been badly injured by a falling tree. Danny managed to pull him from under the tree as the ash continued to fall. Brian was screaming with pain. They sat on a pile of debris, their shirts over their faces, as the ash pelted their backs, first hard and then softer, and then harder again. Sometimes they could hardly see their hands in front of their faces, while at other times the air cleared. The previous night they had camped in a dense old-growth forest. Now they were in a clearing piled high with downed trees. All of the trees were covered in a thick layer of hot gray ash. Brian still had on the long johns that he had worn to bed. Danny had not had time to put on his shoes when he jumped out of the tent, and his feet were now burned nearly as badly as his hands.

They heard Bruce and Sue calling for them, and Danny shouted out a response. Out of the gray fog, Bruce and Sue appeared. Sue looked at Danny. His skin hung from his hands like a burnt marshmallow, and his fingers were fused together. They tried to get Brian to stand, but his hip was broken and he could not walk. For the next hour, they struggled to get Brian to the dilapidated shack on the Green River trail near their campground. It was a short distance away, but they had to lift and roll their friend over dozens of downed trees. They reached the cabin and started discussing what to do. "Don't leave me here to die!" Brian pleaded with them. But the three friends agreed that their odds would be better if they stayed together. Only the front of the shack was intact, so they built a lean-to of logs to cover Brian if more hot ash fell. They assured him that they would return with help.

Bruce, Sue, and Danny made their way across fallen trees to the bridge over the Green River. Danny's and Bruce's pickups were okay, but so many trees had fallen that the pickups could not be driven. The three friends began trudging down Road 2500. This had been second-growth forest, so the trees they had to climb over were smaller. But the ash was burning hot, and Danny had only socks on. They hiked for an hour and a half but made only two miles. Danny was slowing them down. Bruce said that they had to go faster, even if Danny couldn't keep up.

"Can't you stay with me?" Danny begged.

"No," Bruce replied. "We must reach help. Brian'll die. . . . If something happens, get in the water." They left Danny sitting by the river.

Bruce and Sue continued down Road 2500. They wondered if there was anything left of their homes or if the volcano had destroyed all of southwestern Washington. At one point they passed several dozen elk, whose nostrils were plugged with ash. They patted the animals and moved on. Birds lay dead or immobilized in the ash. Sue picked one up and washed it in a puddle, but when she set it in a bush it fell over gasping. I can't save even one little bird, she thought.

Several miles later they came across Grant Christensen, age fifty-nine, who was walking on Road 2500 after driving into the devastation zone earlier that day to try to recover some of his brother's tools from Weyerhaeuser's Camp Baker. Together they began hiking toward safety. Christiansen told them that he'd been a survivor of Guadalcanal in 1942 and had spent two days in a lifeboat before he was rescued. He had a glass eye that he popped out and cleaned whenever it became too ashy.

They were about to stop for the night when they heard a helicopter. They shouted and stomped on the ground, raising a cloud of dust. The helicopter crew saw them and landed nearby. Sue and Bruce had hiked fifteen miles since the volcano erupted.

They told the crew about Brian and Danny. "No one could be alive that far up," said a crewmember.

"We came that far," said Bruce. The crew agreed to search, though darkness was approaching. They flew up the river, but the bridge over the Green River was too small for the Huey to land, and every other potential landing site was covered with logs. They radioed a smaller helicopter nearby, which managed to land on the bridge with its tail hanging over the edge.

But Brian was no longer at the shack. After waiting for an hour, he had grown convinced that he would die if he stayed there. He began to crawl toward Road 2500, trying to keep the weight off his shattered hip. At each downed tree, he crawled underneath if he could. Otherwise he dragged himself up the tree and flopped down the other side. By the time the crew found him that evening and flew him to a hospital in Longview, he had made it only two hundred yards from his starting point.

Farther down the Green River, Danny also had decided that he would die if he did not keep moving. He pushed himself onto his burned feet. Sometimes he walked in the river, where the water made the pain less agonizing. Other times he took the ash-filled road.

Suddenly he heard a voice. "Hey, survivor." It was logger Buzz Smith and his two sons. When the volcano had erupted, they had heard a *crack, crack, crack*, like someone shooting a rifle. Gusts of wind blew through the trees in which they were camping. A black cloud appeared in the sky overhead, and hard objects the size of marbles began falling through the trees. Then it went completely black. They huddled under a sleeping bag and a fallen tree while rocks, mud, and ash fell in the darkness. One boy asked, "Daddy, are we going to live with Jesus?"

"Well, maybe," Smith replied, "but not now."

When it lightened, they began walking down Miners Creek, and then down the Green River. The river was too muddy to drink from, and the boys quickly grew thirsty and tired. Smith told them to think of their trek "like a Hardy Boys story, that we're going to make a very difficult journey in very small steps." As they walked down Road 2500, they saw tracks where birds and insects had walked through the ash. "Look," Smith said, "even birds and hornets are walking. That's how we'll get out."

When they met Danny, Smith put some tennis shoes from his pack on Danny's feet and gave him a fruit roll. The four of them continued walking. Over the next three hours, they walked three more miles to where Road 2500 again crosses the Green River. Smith found a seep of fresh water in the riverbank, and all four of them drank deeply. As Danny drank, Smith was amazed to see water oozing from his neck. "No skin," he later recalled. "He'd been so caked in ash I hadn't noticed this. Now I saw how bad off he was and realized I should have paid more attention."

"You're hurt," he told Danny. "You need to drink more water."

A hundred yards past the bridge, at about seven thirty in the evening, a National Guard Huey flew over them. It landed farther down the road, and crewmen carried Smith's sons to the helicopter while the two men trudged the rest of the way. Danny had walked nearly nine miles on his severely burned feet before the helicopter took off to deliver him to the Longview hospital.

MIKE, LU, BONNIE LU, AND TERRA MOORE

SUNDAY MORNING, LU MOORE WAS PREPARING BREAKFAST FOR HER two daughters and husband as Mike wandered around the campsite taking pictures with his camera. Suddenly she felt as if a giant hand were squeezing her body. Her ears popped, so that she had to swallow to equalize the pressure. Mike was a geology graduate and immediately wondered if it was an eruption. But how could they be feeling the effects way out here on the Green River?

Then Lu noticed a cloud rising above the ridge to the south, and it rapidly grew larger. Mike ran to a nearby clearing, realizing that an eruption was going on and thinking that he could get some photographs of it. But as soon as he got out of the trees, he understood that this was an eruption unlike any that had ever occurred before. The cloud towered above him, its blackness transformed into an immense wall of grays and yellows. The face of the wall was churning, as if driven by an eggbeater. It was one of the most beautiful things he had ever seen, Mike thought.

By the time the blast reached the Green River, it was no longer flowing over every obstacle it encountered. Instead, it was following the contours of the land, like a wave sweeping around rocks. Here the trees knocked down by the blast did not necessarily point away from the volcano. They formed fantastic swirling patterns, with some trees pointing straight back at the source of the blast.

Just south of the Moores' campground is a high point of land known as Black Mountain. The blast swept around either side of the mountain, engulfing the party from Kelso and Longview on one side and Croft and Handy on the other. But because the Miners Creek campsite was already taken by Sue Ruff and her friends, the Moores had camped in the shadow cast by Black Mountain. None of the old-growth trees in which they were camping fell.

As Mike continued to take pictures, Lu quickly took down the tent and threw things into their backpacks. The thunder from the nearby cloud was so loud that they had to shout to each other. The cloud came closer. They decided that their best bet was to take shelter in a dilapidated elk hunter's shack nearby. Within just a few minutes they managed to get all of their gear and the two girls into the shack. Then the ash began to fall. At first it sounded like massive raindrops hitting the roof of the shack. But when Mike stuck his hand outside the door, warm ash and small mud balls bounced off his palm. The valley outside the shack got darker and darker until, suddenly, all of the light was gone. The thunder was so loud, coming every five to ten seconds, that it was impossible for them to hear one another, but they never saw a flash of lightning, so thick was the ash.

For more than an hour, the shack was engulfed by darkness. They covered the baby with blankets and plastic tarps and breathed through dampened socks to keep ash out of their lungs. Bonnie Lu was scared, but Mike and Lu explained what was happening, and she calmed down. The ash was devil's snow, she decided, which made her feel better.

When it became light again and the ash stopped falling so heavily, Mike pushed open the door of the shack and stepped outside. Everything was covered by several inches of gray, powdery material. It was getting lighter to the west. They would be okay, he thought. Lu nursed the baby.

They decided to begin walking the two and a half miles back to their car. Lu strapped the baby into her backpack, wrapping her bald head in a thermal blanket. Bonnie Lu walked alongside in a state of excitement and shock.

After covering about a mile, they began to encounter trees that had fallen across the trail. These blowdowns weren't here the previous day. Why were they here now? they wondered. They veered off the trail to try to get around the newly downed trees. With every step, ash billowed from the ground and nearby vegetation, partially blinding Mike. Lu had to wash out his eyes for fifteen minutes with water from their water bottles before they could continue.

They got past the first few blowdowns. But then they emerged from the trees and beheld an incredible sight. Ahead of them the entire forest had fallen. They thought they might have somehow gotten onto the wrong trail, since they had not encountered anything like this the previous day. But slowly the realization dawned: the volcano had destroyed the mile of old-growth forests between them and their car. They couldn't get over or around the trees with two children, especially given how tired they were. With only three hours of daylight left, they knew they would have to spend another night in the woods.

Mike and Lu were experienced hikers who always carried extra provisions in case of an emergency. They retreated to the standing trees and set up their tent off the trail so they wouldn't be trampled by the elk whose footprints they had seen in the ash. They had dinner with the girls and fell asleep.

Remarkably, they all slept soundly until well after the sun had risen Monday morning. By ten in the morning they were back at the blowdown and ready to start making their way to the car. For the first five hundred feet or so they walked along two downed logs parallel to the path. But then the route got much tougher. They had to climb over eight-to-ten-foot logs piled in haphazard, ash-covered piles. Mike calculated that it would take at least nine hours for them to reach the car.

Shortly before noon, they heard the sounds of a helicopter. Lu was wearing a bright-orange coat, and the helicopter crew had picked out the color against the gray of the ash. The helicopter couldn't land amid the fallen trees, but it lowered a paramedic to the ground. He and the Moores decided to move closer to the river to see if the helicopter could find a place to set down. It took them an hour to cover two hundred yards, but they were able to make their way to an island in the middle of the river.

A small Army Reserve helicopter managed to lower one skid to a gravel bar while hovering above the ash-covered ground. The Moores began to climb in while ash poured through the open door. The pilot was afraid the helicopter would be overloaded. "Leave that damn thing," he shouted, nodding toward Lu's backpack. "We're grossed on weight!" A crewman who had given up his seat so the Moores could be rescued began pulling the pack back out the door.

Lu fought the crewman to keep the pack. "There's a baby in it!" she screamed.

"Okay," said the pilot, "keep the baby."

Three months later, Mike and Lu came back to the Green River. Partly they wanted to look for Mike's $800 Zeiss binoculars, which they found in the ash just where Mike had dropped them. Then they stood and beheld the old-growth forest in which they had camped. As Lu later recalled, it was as if the forest were saying to them, "You've been saved. Now do something about it."

PART 5
THE RESCUES

Station wagon hit by a fallen tree at the edge of the blast zone

THE HELICOPTERS

FOR MORE THAN A MONTH BEFORE THE ERUPTION, DWIGHT REBER, A pilot for the Oregon-based company Columbia Helicopters, had been flying reporters and scientists around Mount St. Helens. The morning of May 18, he had a contract to fly a television crew to film the mountain. When he landed his four-seat Hughes 500C helicopter at the Vancouver airport, the crew rushed out to meet him. "The mountain's blown. Let's go."

Reber had made a promise to the *National Geographic* that if the mountain erupted, he would rescue Reid Blackburn at the Coldwater I observation post. The previous Friday he'd also promised Harry Truman that he would rescue him in the event of an eruption. Reber ran into the airport flight office and phoned the Forest Service. Reber was not allowed to fly to Coldwater I during an eruption, the person on the other end of the line told him. Blackburn had a radio and vehicle. He would have to fend for himself.

For the next hour or so, Reber flew the television crew back and forth just south of the volcano while they filmed the ongoing eruption. Nearby, in a Forest Service spotter plane, was Don Swanson, the geologist who had asked Dave Johnston to fill in for him as Glicken's replacement at Coldwater II on Saturday evening and who was planning to replace Johnston later in the day. Swanson had been in the emergency coordinating center in Vancouver at 8:32 when the seismometer registered a huge earthquake from Mount St. Helens. He dashed to the radio room and called Johnston, but no one answered. An hour later he was flying at an elevation of about 11,000 feet south of the mountain.

Swanson couldn't believe what he was seeing. A thick column of black ash spouted from the top of the mountain and towered many miles above the plane. At one point the wind blew a veil of dust and ash away from the mountain's rim. Swanson recoiled in horror. The top of the volcano was gone. The mountain was at least a thousand feet lower than it had been before the eruption. The ash streamed from an immense crater that extended toward the north, though Swanson couldn't see how far. The forests on either side of the ash cloud had been leveled. Johnston must be dead, he realized.

After Reber dropped off the television crew around 10:45, he took off again for the mountain, this time flying alone. He approached from the south, but the ash cloud was too thick for him to reach Coldwater I that way. He flew west around the cloud and up the north fork of the Toutle River. Visibility dropped, and he flew slowly and close to the ground. But he didn't have enough fuel to get all the way up the Toutle at that speed. He returned to Kelso to refuel.

When he took off again, he flew directly up the valley of the north fork. He looked down and saw not trees and a winding two-lane road, which is what had been there the previous day, but a vast expanse of rocky debris. When the avalanche from the north side of the mountain had hit the ridge on which Johnston was stationed, some of it swept over the ridge but most of it deflected down the north fork. The roadblock on the Spirit Lake Highway, thirteen miles down the valley from Spirit Lake, where the property owners had signed waivers the day before to gain access to their cabins, was completely buried. White geysers spewed from the debris, caused by pockets of water flashing to steam when they encountered hot debris and blast deposits. Reber glanced at his altimeter. He'd flown up this valley so many times that he knew exactly how far he should be above the valley floor. But now the avalanche debris was much closer. It had to be hundreds of feet thick.

Reber found the creek beneath Coldwater I and worked his way up the hillside in about thirty feet of visibility. The formerly forested ridge was covered with ash and craters. To Reber, it looked like an atomic bomb had gone off. Then, out of the corner of his eye, he glimpsed Reid Blackburn's Volvo. Only the roof rose above the ash deposited by the blast cloud. The cameras and tent were gone. He circled around the car trying to see inside. The car was partly filled with ash. No one inside could have survived.

Reber flew back to the valley and continued up the Toutle toward Harry Truman's lodge. He crept along the base of the ridge where Johnston had been stationed. Blue lightning bolts struck all around him. The air smelled strongly of sulfur. If he kept going this way, he'd hit Spirit Lake eventually.

He came to a wide, flat area covered with logs. The air temperature gauge read 140 degrees Fahrenheit; the helicopter strained to stay aloft, but he'd made a promise to Truman. Then, through a hole in the debris, Reber saw open water. This must be Spirit Lake, he thought—or what was left of it. Harry Truman and his lodge were gone.

• • •

Purely by coincidence, two reserve units were conducting training exercises near Mount St. Helens that weekend. The 304th Aerospace Rescue and Recovery Squadron of the US Air Force Reserve was training near Mount Hood about sixty miles south of the volcano. The 116th Armored Cavalry of the Washington Army National Guard was training at the Yakima Firing Center about ninety miles east. The National Guard unit was on call in case anything happened with the volcano while the cabin owners were retrieving their possessions later that day. But no one was prepared for what actually happened. When the response began, it was largely improvised.

The National Guard did not learn that the volcano had exploded until a strange purple cloud appeared in the sky above Yakima. As ash began to fall from the darkening sky, the crews scrambled to their helicopters, with maintenance men wiping ash from the windshields so the pilots could see. But fewer than half of the guard's thirty-two helicopters got away before the ash made further departures impossible.

In Oregon, the 304th did not get orders to send its thirteen-seat Hueys to Mount St. Helens until after ten a.m. More than an hour later, two helicopters made their way up the north fork of the Toutle. "Look at that," one of the crewmen exclaimed. "What the hell happened?" Ahead of them the landscape made an abrupt transition from green to gray. The forest had been flattened. Large trees lay like jackstraw across the Spirit Lake Highway.

About two hundred yards inside the boundary of downed timber, a Ford Torino station wagon had been struck by a falling fir. It had been one of about a half dozen cars, trucks, and motorhomes at the roadblock on Spirit Lake Highway, a few miles farther up the valley, when the volcano erupted. For a few minutes after the eruption, the people at the roadblock had stood outside their vehicles and watched as a dark semicircular cloud rose in the east. Then they could see it coming down the valley toward them. "Time to get the hell out of here," someone at the roadblock shouted. A man and woman jumped into the station wagon and sped away.

Photographer Kathie O'Keefe and Gil Baker were also at the roadblock when the mountain erupted. As the blast cloud approached, they ran to their green Oldsmobile Toronado and fled. Soon Baker was doing 100 miles an hour on the straightaways, yet still the cloud was gaining on them. Projections of the cloud passed them on either side and overhead, so that it felt as if they were driving in the midst of a gigantic black shell. Glancing out the windows, Baker saw that the cloud looked like boiling oil, with immense black iridescent bubbles. Two miles below the roadblock they caught up with a station wagon.

"He's only doing 80," Baker shouted. "What'll I do?"

"Pass him," O'Keefe said.

They passed on a blind curve. If anyone had been coming the other way, they could not have avoided a collision.

In his rear-view mirror, Baker saw the black cloud envelop the station wagon. But he and O'Keefe were finally starting to outrun the blast.

They didn't stop until they reached Interstate 5. There they heard that light ash was falling in Morton.

"Listen to this! All we escaped is a little ash."

"For that we drove like maniacs. Risked our lives."

About noon, the two Hueys from the 304th hovered over the station wagon and tried to determine if anyone was still alive. The headlights were on, but there were no footsteps around the car. Next to the car a man lay motionless in the ash. One of the pilots tried to hover near the car so that two pararescue jumpers could lower to the ground on the helicopter's 250-foot steel cable. But the rotors stirred up too much dust to see, and the helicopter pulled away. One of the jumpers noticed that the adjacent Toutle River bank was still damp. The pilot flew to the river, and jumpers leapt from a skid into thigh-deep water. They waded to shore and began making their way to the car. It was slow going. They had to climb over immense trees and through hot ash, which puffed up around them with every step. They could hear the mountain grumbling upriver. What if it erupted again?

A half hour after dropping into the river, the jumpers reached the station wagon. Its rear door had been hit by a tree trunk, tearing open a gash that let the scorching blast cloud into the interior. The man next to the car was dead. Inside a woman sat in the passenger seat. The dust on her lips was still damp, so she had lived for a few seconds after the blast cloud hit. When the parajumpers touched her skin, it sloughed from her body.

They looked in the backseat through camping gear, grocery bags, and newspapers for any other passengers but found no one. One of the jumpers grabbed the man's wallet. His name was Fred Rollins. He and his wife, Margery, were retirees from Hawthorne, California, near LAX, who had driven north to see the volcano. A parajumper grabbed Margery's purse to stash Fred's wallet inside. When he opened the purse, it was full of bags of cocaine and wadded-up paper money.

• • •

Most of the helicopter rescues that Sunday took place to the northwest of the mountain, where the blast cloud's effects were most severe and where the truncated danger zones attracted the most observers. But all around the mountain that morning, people had been forced to flee the oncoming blast.

Seven and a half miles west of the summit, Ty and Marianna Kearney were watching the mountain from their Dodge van when the avalanche began. They were part of the Radio Amateur Civil Emergency Service group, and Ty had been talking by ham radio with Gerry Martin when the north flank gave way. "We are leaving the area!" he radioed. But the only way the Kearneys could escape was to drive a half mile down a gravel road pointing directly toward the volcano. Marianna, an artist who later published a book of sketches and poems about Mount St. Helens, remembers that the blast cloud looked like black velvet etched in silver and gray as they drove toward it. They reached the turnoff to the road going south—and to safety—just as the cloud shot overhead. A few miles later they emerged into sunlight. They stopped to look back at the volcano. The scene resembled "a monstrous mural in pastels of swirling grays," Marianna later recalled, with the eruption column "a huge cauliflower, light gray against the deep gray gloom of the rest of the cloud."

Eleven miles northeast of the summit, Gary Rosenquist was one of a half dozen people who had camped at a log landing now known as Bear Meadow on a gravel road with a clear view of the mountain. He had been up all night drinking beer and talking around the fire with two of his fellow campers, Joel and Linda Harvey. Sunday morning he attached his Minolta 35mm camera to a tripod. It had twenty-four shots remaining on a thirty-six-shot roll of color film.

At 8:32, someone at the landing said, "There it goes." Rosenquist ran to his camera and grabbed it with both hands, accidentally turning it slightly to the right. Amazingly, the camera was now centered on the landslide and blast cloud rather than the mountain. From 8:32:47 until 8:33:23 he took twenty-two photographs. They remain the best, most complete, and most detailed visual record of an exploding volcano ever taken.

As the blast cloud hit the ridges between the mountain and the landing, it billowed high into the air, like a gigantic wave breaking against offshore reefs. As the cloud approached, Rosenquist unscrewed the camera, threw the tripod in the backseat of the Harveys' Datsun, and jumped in. They careened four miles east down the logging road and turned north toward Randle. A mile later the blast cloud caught up with them. Rocks began falling on the car, followed by balls of warm, dry mud. Harvey tried to stay on the road but could not see through the windshield. Ash began to rain from the sky, and the air smelled of sulfur. A pickup passed them, and Harvey tried to follow its lights. Slowly they made their way up the road with a caravan of other people who had been camping east of the mountain. Ash fell on them all the way to Randle, twenty miles to the north. By noon, sixty people had gathered in the basement of a Randle church to escape the ashfall.

Five and a half miles south of the summit, Forest Service employee Kathy Anderson was leading a contract crew of about twenty-five who were planting noble fir seedlings in a recently logged area. The previous week she had been repeatedly assured that she and the crew would be safe. Still, each of their trucks contained water, forty-eight meals, and filter masks.

In the bright morning sunshine, Anderson watched a double column of black smoke rise above the mountain's summit. Then an immense gray cloud appeared and flowed northeast, away from them. "Run for your lives," a crew foreman screamed, "we're gonna get burnt up!"

Lightning threaded the ash column, and thunder rolled down the mountainside, but other than that the eruption was silent. One strange aspect of the eruption was that the people near the volcano heard no sound of an explosion. People in the blowdown zone heard the ominous sound of rocks hitting one another and the dull thuds of the trees as they fell, but they did not hear the blast itself. Partly, the heavy, ash-laden blast cloud muffled the sound of the explosion. Also, the sound traveled upward more than outward. High in the atmosphere, the sound of the initial blast encountered temperature inversions that reflected and compressed the sound waves. Many hundreds of miles away—as far as Edmonton, Alberta; Butte, Montana; and Redding, California—people heard loud noises, like sonic booms or rifle shots. The sounds seemed to come from very nearby. Police switchboards lit up as people called to find out what had happened.

Standing at the base of the mountain's south side, Anderson had a choice to make. If she led her crew east, they would have to cross two low valleys that would likely flood. If they went west, they would have to cross the bridge on Swift Creek, which Anderson had been told could be washed out in an eruption, in which case they would be trapped. As an ash cloud billowed down the south side of the volcano toward them, Anderson noticed that the larger ash cloud was blowing east—west seemed like the better way to go. They reached the bridge over Swift Creek just as the ash cloud extended over their heads—the bridge was intact. A crewmember waved each vehicle across and got in the last one. Within a few minutes, they had left the cloud behind.

Other people who were watching the volcano that morning did not get away. A man sitting in a pickup with his wife nine miles from the summit was killed when a rock crashed through the roof of the cab and hit him in the head; his wife died shortly thereafter of ash asphyxiation. The half dozen or so people who stayed too long at the roadblock on Spirit Lake Highway were suffocated and cooked by the ash cloud and then buried under the north-fork avalanche. Several people taking photographs from ridgelines to the west and northwest of the mountain were blown from their viewpoints and found hundreds of yards from their vehicles. In almost every case, the victims died quickly, either from breathing in ash or from the force of the blast.

THE FLOODS

THE BLAST WAS THE MOST IMMEDIATE THREAT TO THE PEOPLE around the mountain, but other dangers followed. Sunday morning, Venus Dergan, age twenty-one, and Roald Reitan, twenty, were camping on the Toutle's south fork, twenty-six miles west of Mount St. Helens. Late that morning, they awoke to a strange sound, like the rumble of distant thunder. When they emerged from their tent, the first thing they noticed was the river. From the placid stream of the previous evening, it had changed into an angry gray swirl, and even as they watched they could see it rising. They looked upstream. Suddenly a structure appeared. It was a wooden railroad trestle floating high on the waves. Roald and Venus ran for their car. The river followed. It surrounded them as they got into the vehicle. "Quick, climb on top," said Roald. As they looked around at the river, they saw that it was filled with logs, de-limbed and cut on both ends, obviously from a logging operation upstream. Their car began to float, and then to tilt. They jumped. Roald straddled a log. Venus landed in the water and disappeared from view. Roald glimpsed a flash of black hair and reached for her, but then she was gone.

When the volcano erupted that morning, water from melting snow and ice poured into the south fork of the Toutle River, which is over a tall ridge from the north fork. The flood swept down-valley and hit

Weyerhaeuser's 12 Road Camp about 10:30. There it carried away trucks, loaders, railroad cars, crummies, buildings—and thousands of logs waiting to be sorted and shipped to sawmills. As on the Chippewa a hundred years earlier, the logs formed an immense battering ram that devastated structures downstream. It took out Weyerhaeuser's railroad bridge across the south fork—that's the trestle Roald saw floating down the river toward them. It swept up riverbanks where the river curved, snapping off trees and adding them to the flow.

The logs were so tightly packed that Roald could remain upright on the one he was riding as it carried him downriver. But then the logs shifted and his right leg was crushed. He struggled to free it but could not. He and Venus were in a killing machine, a crazy tumult of water and logs and mud descending through the forest like a runaway train. He realized that he was going to die. It seemed so unfair. Just a few minutes earlier he and Venus had been lying quietly together, and now she was gone. But then the logs shifted again and his leg was free. At almost the same time he saw Venus's hand in the mud ahead of him. He leapt to another log and reached for her. Twice he grabbed for her arm and pulled, but both times the rolling and bounding logs tore her away. He saw her floating between the ends of two sawed logs. Her face was covered in mud, only her eyes visible. If the logs came together, she would be crushed. Instead, the logs separated, and Roald grabbed her hand. This time he was able to keep hold.

Together, they began pulling themselves across the tumbling logs. The water was thick with mud, like wet concrete, and oddly warm, the temperature of bathwater. Venus's wrist was broken, and she and Roald were bleeding heavily from their wounds. It did not seem possible that they could survive. But about a mile below their campground the river channel widened and the water slowed. Roald and Venus began to claw their way from log to log toward shore. As they neared the edge of the water, they saw that they would need to jump from a log toward the steep bank, not knowing if the water beneath their feet would be shallow enough to stand. "Jump!" shouted Roald, and together they leapt from the log. The water came up to their chests, but their feet held. They struggled ashore. They emerged from the river, thickly caked with mud, onto a dirt road. "We'd been in the river five minutes," Roald later recalled. "It seemed like five hours."

They were on a high point of the road, which was covered by the flood on either side. They had no choice but to climb a steep and brushy slope away from the river, despite their injuries. When Venus looked at an open gash on her arm, she could see bone. On the top of the slope, they made their way toward a bridge, where people stood watching the flood. The people on the bridge heard their yells, and two men waded through the mud to help them. At about 12:30, a Weyerhaeuser helicopter hovered over the clearing where they stood and picked up Venus. A few minutes later, a National Guard pilot plucked Roald from the road. Worried about Roald's condition, the pilot yelled to keep him from going into shock—an old military trick. "What the hell happened?" Roald asked. The pilot turned the helicopter so Roald could see the column of ash rising from Mount St. Helens. "That's what happened."

• • •

The flood on the south fork of the Toutle River was just a warm-up for what was to come. In the hours after the eruption, the avalanche debris that had streamed seventeen miles down the Toutle's north fork did something that no one had expected. The debris was filled with ice blocks, snow, and groundwater from inside the mountain. As the debris settled and the ice melted, the water released from the debris formed into ponds, then trickles, then streams, then a great raging muddy flood. Shortly after noon, a Seattle news crew in a helicopter over the north fork saw a chocolate river of mud flowing over the debris avalanche. "How bad is it?" the Forest Service radio operator asked. "It could take out everything from here to the Columbia," replied the helicopter pilot.

Three miles below the toe of the avalanche, the mudflow hit Weyerhaeuser's Camp Baker at about two p.m. Here the devastation was even worse than at the 12 Road camp. The mud picked up logs, vehicles, and entire buildings and carried them downstream. A Weyerhaeuser pilot flying overhead later recalled, "Camp Baker had [been] built over four decades, thousands of men in and out. Now in three hours most of it [is] gone."

As the mudflow swept down the river, it began to take out road and railroad bridges—eventually twenty-seven bridges would be destroyed. Where the Spirit Lake Highway crosses the Toutle at the Coal Bank Bridge, people gathered on either side to watch logs slamming into the bridge's upstream flank. First the north side of the bridge came loose, pivoting downstream. Then, with a puff of concrete dust, the south end of the bridge gave way. The two-hundred-foot bridge rotated into the river and sank into the mud.

Along the more developed parts of the river, the mudflow crept over riverbanks and onto broad lawns. It surrounded barns, garages, houses. Property owners watched their homes wrench free from their foundations and enter the flow. "Lots of popping and snapping sounding like boards as well as trees told me the flood was taking our house," said one. "Nothing we could do." By the time the flood subsided, nearly two hundred homes had been damaged or destroyed.

With bridges over the rivers destroyed and roads covered with flowing mud, people were stranded. The Hueys began picking up people and flying them to shelters. Some of the Hueys were dangerously overloaded. Yet no one was injured or killed in the rescue operations, despite the risks that were taken. By the end of the day, the helicopter pilots had flown 138 people, 8 dogs, and 1 boa constrictor to safety.

• • •

That afternoon, a National Guard helicopter flew up the north fork after receiving reports of footprints seen in an ash-covered logging road. After forty-five minutes of searching, the helicopter crew spotted the footprints and followed them toward the river. There, on a tree trunk in the middle of the mudflow, was José Diaz, a recent immigrant of Mexican origin, barefoot and severely burned. He was part of a four-man gyppo crew on a Weyerhaeuser contract that had been thinning Douglas fir thirteen miles northwest of the volcano. Sunday morning, Diaz had been resting in a crewmate's pickup—a strict Roman Catholic, he had worked the previous day but would not work on Sunday after the four men camped Saturday night near the worksite. When the volcano erupted, the other three workers were cutting trees downslope from the pickup and did not feel the earthquake. Suddenly Diaz came running through the woods. *"El volcán esta explotando,"* he was shouting. He ran right past them and fell into the tangle of trees they had just cut.

Trees upslope from the loggers began to snap off. Hard pellets flew through the falling woods like bullets. A dark cloud descended on them, and they all fell to the ground. As one later recalled, "It got hot right away, then scorching hot and impossible to breathe. The air had no oxygen, like being trapped underwater. I gasped for breath for a minute, and the inside of my throat got very hot. I felt I was being burned, thought even I was being covered by lava. I was being cremated, the pain unbearable."

After a few minutes, the air lightened. All the trees were down and covered in ash, but none had fallen on the four men. They pulled themselves to their feet. They were far enough away not to die immediately in the blast, but all four were badly burned. They could barely breathe in the ashy air, but they knew that they had to move if they were going to live. They climbed over the downed trees to the road and walked back to the truck, where they waited for the falling ash to diminish. Then they began walking back down the road toward Camp Baker. They walked three miles, at which point they found themselves trapped by a landslide across the road.

They argued about what to do. One man headed down a side road toward the river—he later would be found, dead, suspended in a hemlock tree he had climbed to escape the mudflow. A few minutes later, Diaz also left the two other men and made his way toward the river. (The two remaining loggers were picked up several hours later by a National Guard helicopter. One lived; the other died from his burns.) Diaz reached the Toutle and climbed onto a log where he could be seen, but the river rapidly rose around him. He was trapped a hundred yards from shore when the rescue helicopter appeared.

The log on which Diaz crouched was bobbing in the mudflow. If it rolled, he would be killed. But the helicopter could not land in the mudflow to rescue him. The pilot decided that he would have to hover over the log while his crew pulled the man to safety. But Hueys build up a static charge when they fly. When the flight surgeon reached out the door for Diaz's hand, a jolt of static electricity knocked the surgeon onto his back. The helicopter retreated and tried again. The pilot touched one ski to the tree trunk to ground the machine, and the flight surgeon dragged the logger inside. Two weeks later, after saying almost nothing to his rescuers or the people in the hospital, Diaz died of his burns.

• • •

In the early evening, the mudflow reached the bridge over I-5. State troopers had stopped traffic on the highway as the flood approached. Now logs began to ram the bridge's supports over and over, its metal trusses vibrating with each impact. But the bridge held, and gradually the logjam passed underneath.

Even before the flood subsided, enterprising boat owners saw an opportunity. They rushed onto the river and began to corral the logs from Weyerhaeuser's upstream camps, each of which was worth several hundred dollars. One entrepreneur used a crane to pluck choice timber from the river and deposit it onto three barges. Weeks later, Weyerhaeuser began suing riverside property owners to get its logs back.

The flood carried immense amounts of mud and gravel that came to rest on the beds of the Toutle, Cowlitz, and Columbia Rivers. Channels became so clogged that boats could not leave their moorings and future flooding became inevitable. Oceangoing cargo ships were stranded for weeks in Portland, fifty miles south of the volcano, while dredging crews labored to clear a channel through the Columbia.

THE ASHFALL

THE FOURTH COMPONENT OF THE ERUPTION—BESIDES THE AVAlanche, the blast, and the floods—was the ashfall. For more than a day, the magma chamber inside Mount St. Helens pumped ash into the atmosphere—altogether a quarter of a cubic mile of pumice and ash, like talcum powder but grittier. Initially it rose to more than 60,000 feet—twice the altitude of a commercial airliner. Then it caught in the westerlies that blow over North America and began to drift east. Over the next two weeks, the ash cloud would travel all the way around the world.

Several commercial pilots flying routes to and from Seattle that morning witnessed the rising column of ash. A Northwest pilot flying a quarter-full DC-10 from Seattle to Washington, DC, diverted south of Mount Rainier to give his passengers a view of the erupting volcano. The ash cloud formed a huge east-west curtain to their south, with lightning bolts running around the cloud's edges. The jet's engineer radioed the tower at McChord Air Force Base south of Tacoma, "You see the smoke to the south?"

"It's the military at Yakima firing range."

"It's Mount St. Helens, and it's huge. People will need help!"

On Mount Adams, thirty miles to the east of the volcano, climbers watched in astonishment as the mountain erupted. As the ash cloud approached, sparks began to fly from ice axes, knives, and anything else made of metal. Ash clouds contain powerful electric charges caused by ash particles rubbing together, in the same way that rubbing balloons together generates static electricity. As a result, volcanic eruptions almost always contain intense displays of lightning. On Mount Adams, a young climber with braces felt the electricity play across his teeth. Around the climbers, charred wood, pinecones, and branches as large as a man's arm fell from the sky, followed by pumice, mud balls, and ash. The climbers made their way back down the mountain in a hazy twilight, sometimes dropping to their knees to feel for their previous tracks in the snow.

• • •

In an era before cell phones, twenty-four-hour news cycles, and the Internet, news was slow to reach people downwind that Mount St. Helens had exploded. In eastern Washington, people going to church that Sunday morning looked to the west and thought that a violent thunderstorm was approaching. The ash cloud looked like an immense blanket being dragged over the sky from west to east. Great bulbous projections, like mammatus clouds but larger and darker, hung from the bottom of the cloud. A bolt of lightning would open up a hole in the cloud, and then the hole would quickly close.

Ash began to sift from the sky. It made a hissing sound as it fell, like air leaking from a tire. In the small eastern Washington town of Othello, 150 miles from the volcano, two inches of ash covered everything—crops, cars, driveways, roofs, roads, golf courses. Students at Washington State University in Pullman, on the border with Idaho, made emergency runs to convenience stores to stock up on beer. People in Spokane ventured into the semidarkness to collect the ash in jars as keepsakes. Daylight turned to darkness, so that streetlights switched on and formed glowing orbs in the falling ash.

People driving on the highways and two-lane roads of eastern Washington gradually began to hear on the radio that Mount St. Helens had erupted, though usually not before the ash started falling. Ash billowing across the roadway made it impossible to drive. They began to stop in small towns and highway rest stops, where many would be stranded for days. Radio announcers reported that people should not breathe the ash or wash it off with water because it would turn to acid, which turned out to be incorrect information but further panicked those stuck in the ashfall. No one knew how much ash would fall or when it would stop.

• • •

The next morning dawned clear and bright. A layer of ash as much as three inches thick stretched from the volcano through eastern Washington, Idaho, Montana, and Wyoming.

Even a year later, ash billowing up behind moving cars was contributing to fatal accidents.

THE DEAD

WHEN THE FOUR MEMBERS OF THE MOORE FAMILY WERE HOISTED from the Green River Monday afternoon, they were the last people to be rescued from the devastated area. After that, the rescuers turned to locating and removing bodies.

On Monday, a photographer flying over a ridge west of the volcano shot a picture of a body lying lifeless in the bed of a pickup. That evening in Portland, the photographer stood next to the darkroom technician as the print began to appear in the developer. "This guy looks awfully young," said the technician. The photographer hesitated, but a few minutes later he sent out the photograph over the Associated Press wire. When the photograph was published in newspapers the next day, the boy's grandparents instantly recognized their eleven-year-old grandson.

Tuesday morning, Dwight Reber flew a *National Geographic* writer and a photographer from the *Columbian* to Coldwater I to show them Blackburn's buried Volvo. He obviously had not survived. But Reber still wasn't able to set the helicopter down. Not until Thursday was a helicopter able to land so that Blackburn's body could be identified. Above the ash, his face was still recognizable. But below the ash his body had burned away.

A forest ranger walking south toward Ryan Lake found Clyde Croft's body in his sleeping bag by the side of the road. Rescuers followed his footsteps to the pickup and horse trailer, and then to the campsite by the Green River, reconstructing the last few hours of Croft's life. At the campsite they found the bodies of Al Handy and the two horses.

On Thursday morning, Sue Ruff and Bruce Nelson were on the *Today* show to talk about their escape from the volcano, after which they had lunch with their interviewer, David Burrington. They told him that they had been trying to get the rescuers to search for their friends, Karen Varner and Terry Crall, but that their requests were being ignored. Burrington chartered a helicopter, put Ruff, Nelson, and a camera crew on board, and headed for the campsite. But the helicopter was ordered to land at the operations center in Toledo before it could head up the Green River. For five hours, they waited for permission to fly to the camp. Finally, Burrington and Nelson, followed by a cameraman, cornered two search officials. "If you don't let these kids look for their friends," Burrington said, "we'll put you on national television and make you look like the assholes that you are."

Shortly thereafter an army helicopter and two National Guard choppers took off and flew Nelson and a rescue party (there was no room for Ruff) up the Green River to the bridge where Bruce and Danny had parked their pickups. As Nelson and the rescuers struggled over downed trees toward the campsite, they saw Tye's three puppies running around in the ash. A small tree had pinned Tye, but when they cut away the tree, she rose and walked. The dogs were alive. Could Terry and Karen have lived too?

They sawed away more trees and found the tent. Terry was lying on his right side, his arms around Karen. They both had been crushed by the tree that had fallen on their tent. He was wearing jeans and a work shirt. She wore the long underwear in which she had slept. An air force reserve helicopter cabled down a basket and lifted the body bags into the sky. Nelson put the puppies in his pack and hiked with Tye back down Road 2500.

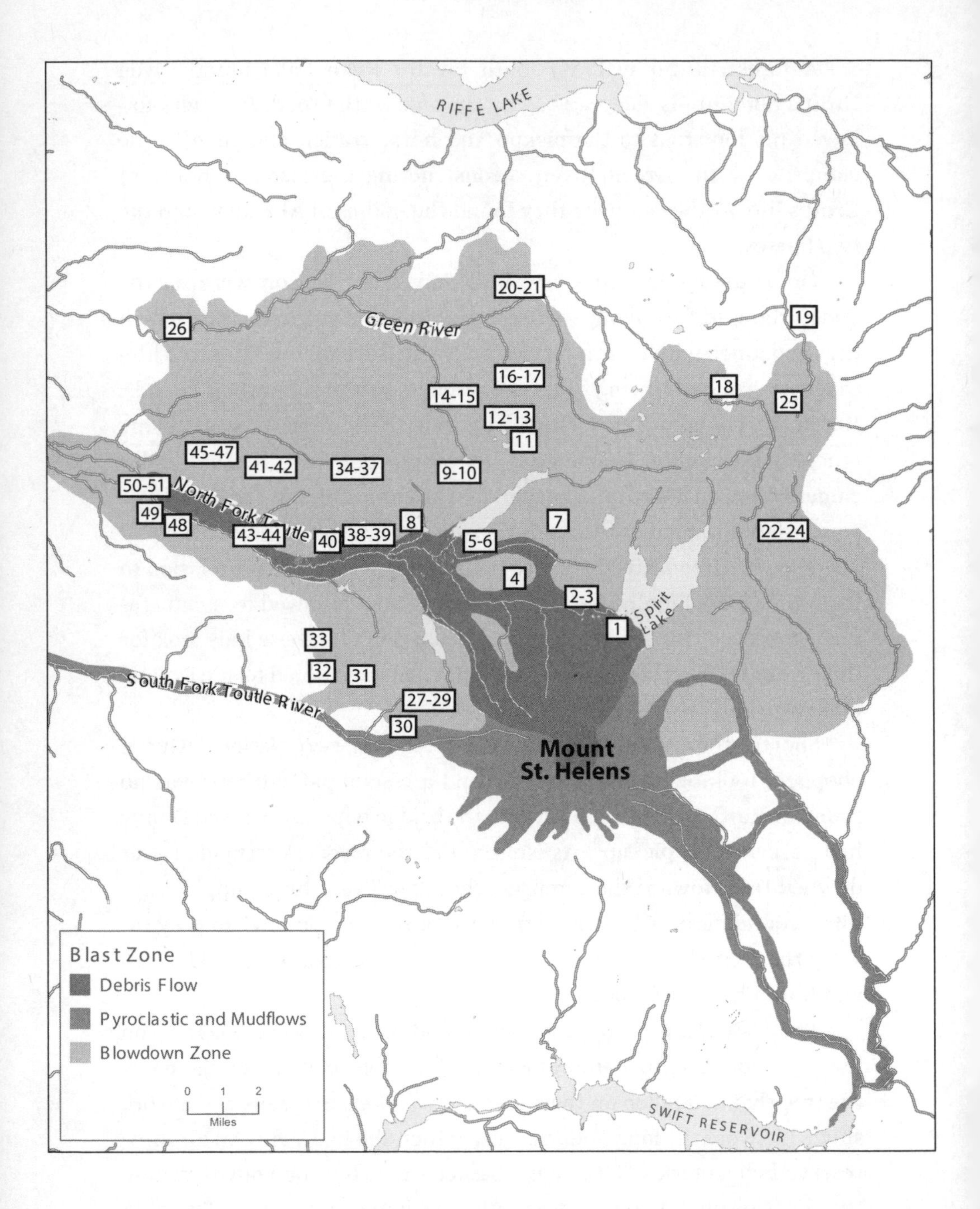

RIFFE LAKE
Green River
20-21
19
26
16-17
18
14-15
25
12-13
11
45-47
41-42
34-37
9-10
50-51
North Fork Toutle
49
48
8
7
43-44
40
38-39
5-6
22-24
4
2-3
Spirit Lake
1
33
32
31
South Fork Toutle River
27-29
30
Mount St. Helens
Blast Zone
Debris Flow
Pyroclastic and Mudflows
Blowdown Zone
0
1
2
Miles
SWIFT RESERVOIR

PEOPLE KILLED BY THE ERUPTION OF MOUNT ST. HELENS

1. Harry Truman
2. Bob Kasewater
3. Beverly Wetherald
4. Dave Johnston
5. Jim Pluard
6. Kathleen Pluard
7. Gerry Martin
8. Reid Blackburn
9. John Killian
10. Christy Killian
11. Robert Lynds
12. William Parker
13. Jean Parker
14. Wally Bowers
15. Tom Gadwa
16. Jerome Moore
17. Shirley Moore
18. Al Handy
19. Clyde Croft
20. Terry Crall
21. Karen Varner
22. Donald Parker
23. Richard Parker
24. Natalie Parker
25. Ronald Conner
26. Keith Moore
27. Day Bradley Karr
28. Day Andrew Karr
29. Michael Murray Karr
30. Robert Landsburg
31. James Fitzgerald
32. Donald Selby
33. Paul Schmidt
34. Ronald Seibold
35. Barbara Seibold
36. Kevin Morris
37. Michelle Morris
38. Arlene Edwards
39. Jolene Edwards
40. Bruce Faddis
41. Harold Kirkpatrick
42. Joyce Kirkpatrick
43. Edward Murphy
44. Eleanor Murphy
45. Leonty Skorohodoff
46. Evlanty Sharipoff
47. José Diaz
48. Joel Colten
49. Klaus Zimmerman
50. Fred Rollins
51. Margery Rollins
52. Ellen Dill
53. William Dill
54. James Tute
55. Velvetia Tute
56. Paul Hiatt
57. Dale Thayer

Note: The locations of Ellen and William Dill and James and Velvetia Tute at the time of the eruption are unknown. The identities of Paul Hiatt and Dale Thayer remain uncertain, though their names appeared on early lists of missing people. Several additional people may have died in the eruption, but their names did not appear on the final official list of the dead.

• • •

On Wednesday evening, President Carter arrived from Washington, DC, to view the devastation. As many presidents have learned the hard way, visiting disaster sites is tricky. A president, his face framed in the window of a helicopter or truck convoy, can look powerless in the face of an overwhelming disaster. A president can hug the bereaved, but offering such solace only makes a president look human, not presidential.

By that time, Carter's reelection campaign was in trouble. The economy, in poor shape throughout his term, had swung back into recession in the first half of 1980. In the primaries that spring, Carter was being challenged from the left by Sen. Ted Kennedy, while a third-party candidate named John Anderson was threatening to siphon the votes of disenchanted Democrats away from him. Following the failed April 25 rescue of the hostages, Iranian mobs were burning effigies of Carter in the streets of Tehran. Carter might try to use the volcano to take people's minds off their troubles, but the scenes of death and desolation just made people more worried and depressed.

Traveling with Carter from DC were Sen. Warren Magnuson of Washington State, then the powerful chair of the Senate Appropriations Committee, and Sen. Mark Hatfield from Oregon, who would succeed Magnuson as Appropriations Committee chair the following year, along with members of the president's cabinet and Carter's science adviser, geophysicist Frank Press. They were greeted at the airport by Governor Ray, Governor Vic Atiyeh of Oregon, and representative Don Bonker, whose district included the area west of the volcano. Then the entire entourage made its way across the Columbia River to Forest Service headquarters in Vancouver, where the president received a briefing from Rocky Crandell, Bob Tokarczyk, and other officials. Crandell warned that the volcano could erupt explosively again, especially once it built up a new cone of eruptive material. Carter asked if the power released by the volcano that Sunday was in the fifty-megaton range. Crandell responded that it was between one and ten megatons—still greater than anything in the American nuclear arsenal.

Finally Governor Ray could no longer keep quiet. "This is all very interesting . . . but the top priority is the people," she said, adding that she was concerned "not so much for the people who were killed—they are gone now."

"What do you need specifically?" Carter asked.

Ray spelled it out for the flustered president: "M-O-N-E-Y."

Carter objected that he had promptly declared the state a disaster area after the eruption. Magnuson, clearly irritated with the governor, cut in: "We're in the red, and the states are in the black. . . . We are trying to balance the budget."

"We understand," Ray retorted, "but we have a problem."

Magnuson said that the federal government could pay for immediate disaster assistance, "but in the long run, I don't know."

Afterward, Bonker said that he was satisfied with the promises made by Carter and Magnuson. But he too was incensed at Ray's behavior. "Don't ask me about the governor," he told a reporter.

The next morning, dressed in a suit, a white raincoat, and army boots, Carter flew toward the volcano in a Marine Corps helicopter. In the helicopter, Governor Ray continued to rant about the state's need for federal funds. Carter turned to Tokarczyk, who was sitting next to him. "My headphones don't work," he said, pleased not to have to listen to Ray's tirades. He glanced out the window. "I'm amazed at all this destruction," he said.

"We're not there yet, Mr. President," said Tokarczyk. "Those are clearcuts."

The mountain and Spirit Lake were once again shrouded in clouds, so the president flew over the avalanche debris in the Toutle River valley before turning back. "I've never seen or heard of anything like this before," he said in a post-flight interview. "Somebody said it looked like a moonscape, but the moon looks like a golf course compared to what's up there. . . . The ash is several hundred feet deep. There are tremendous clouds of steam coming up. There are enormous icebergs, big as a mobile home. A lot of them are melting, and as the icebergs melt . . . the ash caves in and creates enormous craters. There are a few fires about, on the edge of the ash flow, where logs are still exposed. . . . It's an unbelievable sight."

• • •

Even before she flew up the Toutle River with the president, Governor Ray was trying to deflect blame to the victims. "Many people chose to remain" near the mountain despite repeated warnings, she said in a press conference the Tuesday after the eruption. "We cannot be responsible. . . . It's a free country."

President Carter and his aides took their lead from Ray. "One of the reasons for the loss of life that has occurred is that tourists and other interested people, curious people, refused to comply with the directives issued by the governor, by the local sheriff, the State Patrol, and others," Carter said in an interview when he was in Portland. "They slipped around highway barricades and into the dangerous area when it was well known to be dangerous."

This pass-the-buck fabrication has proven extremely persistent. Almost a year after the eruption, *National Geographic*, despite its legions of fact-checkers, wrote in a caption to a feature article on the days leading up to the eruption: "Fearing the worst, state officials barred entry to a 20-mile 'red zone,' its heart mapped at right, except for critical personnel such as geologists, who sought to sniff, feel, and listen to the mountain through sensitive devices. Without restrictions, thousands of sightseers would have died."

At least the *Geographic* got the last sentence right. Without the roadblock on the Spirit Lake Highway and the red and blue zones, many more people would have died. But in retrospect, the restrictions were pitifully inadequate. Today, a visitor to the mountain standing where Dave Johnston stood, or where Reid Blackburn was taking photographs, or where Harry Truman drank his last whiskey and Coke, has to wonder what they were thinking. The mountain is huge. Its blown-out entrails extend miles from the volcano. How could anyone stand there in the weeks before the eruption and not think they were in danger?

The relatively low number of people killed that morning was mostly a matter of luck. Fewer people were around the volcano Saturday night and Sunday morning than at any other time that weekend. If the volcano had erupted Saturday afternoon or Sunday afternoon, the death toll would have been several times higher. And if it had erupted during a weekday, hundreds of loggers would have been killed. Even today, people throughout southwestern Washington will say, "If it had been the week before or the week after, it would have been me."

Since the eruption, Mount St. Helens's victims have been blamed many times for their own deaths. Even in the Pacific Northwest, most people still think that the victims were there illegally, that they went around roadblocks or otherwise broke the law to get where they were. That belief is the product of a carefully fabricated lie. Dixy Lee Ray and other public officials were unwilling to take any blame for the disaster. They stuck steadfastly to the story that the victims had been warned and shouldn't have been where they were.

But blaming the dead misrepresents the facts and insults the memory of the victims. It doesn't even make sense. Was ham radio operator Gerry Martin, perched above South Coldwater Creek to warn others of an impending eruption, breaking the law? Would the Killians break the law to go camping at Fawn Lake when they couldn't even see the volcano from there? If the volcano had blown up twenty-four hours later, would the hundreds of Weyerhaeuser loggers who would have died be responsible for their own deaths?

No one was acting illegally because there was no law to break. None of the people camping around Mount St. Helens that morning went around the roadblock on the Spirit Lake Highway. And because the red and blue zones ended at Weyerhaeuser's property lines, police officers did not even try to keep people off Weyerhaeuser land. The extension of the blue zone sitting on Dixy Lee Ray's desk Sunday morning would have helped get people out of the danger zones, but who knows if she would have signed it. She later said that she didn't believe in the blue zone: "It's like, you can't be half pregnant. . . . You have a place you say people should stay out of, but not a sort of half-assed place, where they might and they might not. I never did accept the concept of a blue zone." Only three people were in the red zone, and two of them—Bob Kaseweter and Beverly Wetherald—had permission to be there. In fact, the only person who broke the law was the one person who emerged from the disaster with his reputation relatively enhanced: Harry Truman.

Recognition, and a measure of forgiveness, have been slow in coming to the fifty-seven victims of the Mount St. Helens eruption. But today, near the spot where Dave Johnston was watching the volcano that sunny Sunday morning, a stark gray granite slab overlooks the avalanche debris above the north fork of the Toutle River. It says IN MEMORY OF THOSE WHO LOST THEIR LIVES IN THE MAY 18, 1980 ERUPTION OF MOUNT ST. HELENS. Then it lists their names. Beyond, the gaping maw of the blasted-open volcano lies bleached in the sun.

PART 6
THE MONUMENT

Susan Saul hiking above Spirit Lake in the Mount St. Helens National Volcanic Monument

THE AFTERMATH

WHEN MOUNT ST. HELENS ERUPTED ON SUNDAY, MAY 18, 1980, Susan Saul was helping to present an environmental education workshop for teachers at the Malheur National Wildlife Refuge in southeastern Oregon. The isolated refuge had no radio or television reception. All that day, she heard nothing about what had happened.

About noon on Monday she got in her car and began driving home. She turned on the radio and heard that the governor of Idaho had declared the northern part of the state a disaster area. What could have happened in northern Idaho? She drove farther and heard that the disaster was ashfall from Mount St. Helens. That was odd. The volcano had never emitted that much ash before. As she approached Bend, she began to hear more detailed reports about a massive eruption. But it was not until she reached home that evening and began reading the newspapers piled up on her doorstep that she learned exactly what had happened. Spirit Lake was gone. The mountain was 1,300 feet lower than it had been before. Dozens of people were dead. Almost everything she had been fighting to preserve had been destroyed.

The newspapers said that a huge mud dam created by the eruption could collapse at any time, releasing the waters of Spirit Lake down the Toutle and flooding downstream communities. Longview was under a flash-flood alert. Residents were supposed to put clothes and a sleeping bag in their cars so they could evacuate immediately if told to do so.

Saul knew she would never be able to get to sleep at her house because she would worry all night about missing an evacuation order. But she was so tired from driving all day and trying to get news on the car radio that she desperately needed to sleep. She put enough clothes to last for several days in the backseat of her car, and she and her cat drove to a friend's house on a hill above Longview.

The next day at work, she and her coworkers at the Fish and Wildlife Service listened to the radio all day in case the call to evacuate came. They also had to conserve water, because the intakes of the Longview filtration plant were clogged by mud and the water in the city's reservoirs was dropping rapidly. She stayed at her friend's house again that evening. Finally on Wednesday they heard that the mud dam holding back Spirit Lake was expected to hold and that they would have five to six hours of warning if it failed. That night, for the first time since arriving back in Longview, she slept in her own bed.

Thursday morning she heard on the radio that President Carter was going to pass through Longview on his way to visit a refugee center at a nearby middle school. She and a friend at work walked a few blocks from the office to watch the motorcade go by. Carter was just another gray-haired man in a raincoat riding in a large green station wagon. She laughed at the motorcade and at herself, and for the first time since her return she began to feel normal. After getting home that evening, she worked in her garden.

Saul knew how close she and her hiking party had come to being killed. As Harold Deery told her that week, "We were playing Russian roulette when we went on that hike." She went back and read the chapter on Mount St. Helens in Stephen Harris's book *Fire Mountains of the West*. They had all been too sanguine. Mount St. Helens was one of the most violent of Cascade volcanoes. It had erupted explosively in the past and would do so again. The area around the mountain had been wiped clean many times before. Saul took out her slides of the mountain and flashed the photographs on the wall of her living room. Spirit Lake would never again look the way she remembered it. Someday its shores would be forested, and streams would cascade from the surrounding highlands into the clear blue water, but not in her lifetime.

On Sunday morning, one week after the eruption, Saul looked out the window of her house and saw that everything was covered with ash. The mountain had erupted again overnight while the wind was blowing from the east. A gray rain mixed with the ash and dripped from branches and leaves. Maybe this is the way it would be from now on living near the volcano.

• • •

On June 5, Saul, Noel McRae, and the leaders of several groups that had supported the work of the Mount St. Helens Protective Association met to consider what to do next. Should they give up their plans to establish a monument around the now-devastated volcano? If they moved forward, what did they want to protect? They decided to spend the summer gathering information, touring the area, and meeting with scientists and government officials. In the fall, they would meet again to make a decision.

The first problem they faced was seeing exactly what the volcano had done. Access to the red zone was forbidden to those without special permission, and newspaper photographs gave only a piecemeal impression. Their best bet to get into the red zone, they decided, was to get to know the scientists who had descended on the mountain to study the aftermath of the explosion. The scientists were still mourning the loss of Dave Johnston, but that didn't stop them from taking advantage of a unique opportunity. In the walls of the ripped-open volcano, geologists could read the layers of deposition and destruction like the pages of a book. The landslide had created a strange hummocky landscape that bore an unexpected resemblance to landforms near other volcanoes around the world. Most interesting of all, the eruption had scoured the land of living things in a vast tract north of the volcano. Biologists were eager to watch life recolonize the blast zone to learn how the earth can recover from even the most severe devastation.

The scientists with whom Saul and her associates talked clearly wanted the area around Mount St. Helens to be preserved. There was nothing like it anywhere in the world. Scientists could study the area for years and still have much more to do. Maybe the protective association could build a case for preserving the area as a giant laboratory.

Then the association got a break. The US Army Corps of Engineers wanted to start building dams to hold back sediment before the fall rains started. The only way to do that would be to avoid lawsuits from conservation groups that wanted to slow progress. To show the leaders of the conservation groups how badly the dams were needed, corps officials took them on a helicopter tour of the blast zone. On August 12, the group boarded National Guard helicopters at the Kelso airport, flew above Spirit Lake and the north fork of the Toutle, and landed on the bank of the south fork.

The valley was almost unrecognizable. Saul and the other conservation leaders walked on hard-packed mud that had flowed from the flanks of the mountain. The ground was pockmarked by craters where melting pieces of ice had erupted in geysers of steam. The volcano loomed above them, gray and silent.

They were lucky for another reason that day. Also on the helicopter tour was a young legislative aide to Congressman Bonker named Jim Van Nostrand. Saul and the others talked to him about their past efforts to protect the area around the volcano. Van Nostrand was interested. This was something his boss could get behind. Preserving the area would create new jobs, replacing at least some of those lost when logging operations and mills shut down after the eruption. It could bring tourists back to the Third District, no longer to swim and fish in Spirit Lake but to see the remnants of an exploded volcano.

In October the conservation group leaders convinced the Forest Service to give them a tour of the Green River valley. They drove up the Forest Service road from Randle and parked at Ryan Lake. The trees still lay thickly scattered around the clearing where Handy and Croft had parked their borrowed horse trailer two days before the eruption. But when they looked down the Green River valley, they could see the intact old-growth forest in which the Moores had camped.

They were working against time. The previous month, Weyerhaeuser and other logging companies had begun salvaging trees from the blowdown zone. The companies argued that if they did not remove the downed trees quickly, the wood would rot, or be consumed by insects, or burn in a fire. The Forest Service agreed and already was signing emergency contracts to salvage the trees on federal land. Weyerhaeuser planned to have more than a thousand men in the woods by the following spring removing the 68,000 acres of fallen timber on Weyerhaeuser land. Within two years the job would be done, and new trees could be planted to prepare another crop for harvest.

Saul and the other conservationists knew that once the trees were removed it would be much harder to protect a given parcel of land. Salvaged areas would no longer be of interest to biologists or to tourists who wanted to see the effects of the eruption. Weyerhaeuser would be back in control, just as it had been before the eruption. As Saul told a reporter, "Land management decisions are being made with the bulldozer and chainsaw."

For the loggers, the salvage was dangerous, depressing, and dirty work. Felling, yarding, and loading the ash-covered timber raised clouds of choking dust. The men had to wear hospital masks despite the heat. Cutters could take down only two trees before they had to resharpen their dulled chainsaws. Each crew had to be in radio contact with a base station, which would be notified by the Forest Service if the mountain showed any sign of activity. Over the course of that summer the volcano erupted four more times—on May 25, June 12, July 22, and August 7—releasing less ash than on May 18 but far more than during the phreatic eruptions of the spring. Still, the loggers were constantly worried. "It was like going to the land of Mordor," said one logger. "It wasn't just the volcano that gave us a feeling of dread—the whole gray, lifeless landscape made us apprehensive. . . . Whenever a job came up in the Red Zone, we'd say, 'Oh, no, back to Mordor,' and think of all kinds of excuses not to go to work that day." Despite the loggers' discomfort, by October Weyerhaeuser was running three hundred fully loaded trucks out of the blast zone each day.

• • •

On November 4, 1980, Ronald Reagan soundly defeated Jimmy Carter for the presidency. Carter won only six states: Georgia, Hawaii, Maryland, Minnesota, Rhode Island, and West Virginia. It was the worst defeat for a sitting president since Herbert Hoover lost to Franklin Roosevelt in 1932.

Reagan's candidacy coincided with a pivotal time for the US environmental movement. Conservation groups had grown dramatically during the 1970s. National organizations like the Sierra Club, the Audubon Society, and the Wilderness Society pushed for federal legislation, while thousands of local groups like the Mount St. Helens Protective Association worked on specific issues. But the success of the movement had contributed to a conservative backlash. In the late 1970s, a group of angry westerners formed what they called the Sagebrush Rebellion to oppose what they considered unreasonable restrictions on the use of publicly owned lands for grazing, mining, and logging. They urged that land controlled by the Forest Service, the Bureau of Land Management, and other federal agencies be returned to the states, so that it could be managed in a more economically productive manner. The movement shied away from appealing to states' rights, still associated with slavery and the Civil War. Rather, it drew its imagery and rhetoric from the Revolutionary War. "We're talking about issues that haven't come up since the Boston Tea Party," said one spokesman for the rebellion.

Actually, very similar movements had occurred twice before in the twentieth century. Between 1907 and 1915, right after Roosevelt and Pinchot expanded the national forests, six conferences in western states gave ranchers and others an opportunity to demand changes in the federal management of rangelands. But the public was unsympathetic to ranchers' complaints about having to pay for grazing their livestock in the national forests, and the movement faded away. Again, in the 1940s, congressional hearings throughout the West provided stockmen with a forum to object to possible reductions in grazing to improve the conditions of publicly owned rangeland, which was being destroyed by privately owned livestock. But the ranchers' proposal to buy rangelands from the federal government for prices ranging from nine cents to $2.80 per acre went nowhere.

During his campaign, Reagan had offered his support to the rebellion. As he wrote in a telegram shortly after his election, "I renew my pledge to work toward a 'sagebrush solution.' My administration will work to insure that the states have an equitable share of public lands and their natural resources." The underlying message to groups like the Mount St. Helens Protective Association was clear: Once Reagan was in office in January, it would be much harder to protect the area around Mount St. Helens from logging, mining, and other private uses.

There was another way to protect the land if the conservation groups moved quickly. Many times in the past, presidents had used executive orders to designate areas of special scenic or historical significance as national monuments, which do not have to be approved by Congress. Carter had already done so in 1978, when he designated 56 million acres in Alaska as national monuments after Congress refused to pass an Alaska lands bill. If Carter created a national monument around Mount St. Helens before he left office, the association would not have to fight an uphill battle against Reagan and his Sagebrush Rebellion allies. Conservationists started lobbying the White House to act, and environmentalists within the White House added their voice to the cause.

Carter was terribly distracted. The hostage crisis was coming to a head, with Algeria negotiating with Iran for the hostages' release. Both inflation and budget deficits were at record high levels, and the poverty rate was growing. The prime interest rate reached 21.5 percent that December, the highest it had ever been. The Alaska National Interest Lands Conservation Act finally was moving through Congress, which would affect an area hundreds of times larger than the area around Mount St. Helens. The remote part of Washington State that he had visited in the spring, with its prickly governor, must have seemed an afterthought to the president.

In the last week of 1980, the president held a meeting in the White House to discuss the possibility of designating the area around Mount St. Helens as a national monument. No representative of any conservation group attended the meeting. But two representatives of timber owners—a forest engineer from Weyerhaeuser and a representative of the Washington State Department of Natural Resources—flew to Washington, DC, to be there. After the meeting, according to an internal Forest Service communication, Mount St. Helens was withdrawn as a possible national monument.

On January 20, 1981, just as President Reagan finished his first inaugural address, Iran released the hostages into US custody after 444 days in captivity. Carter's presidency, and the hostage crisis that had helped end it, were over.

THE PROPOSALS

FIVE DAYS BEFORE RONALD REAGAN WAS SWORN IN AS PRESIDENT, the Mount St. Helens Protective Association held a news conference to announce a new proposal for preserving the area around the mountain. The association and its allies had decided to go big. As the Sierra Club's Charlie Raines, who helped fashion the proposal, later described it, "We have an incredible opportunity to protect one of the world's great wonders here, with unsurpassed value for scientists, educators, and the general public. Let us proceed with vision, leaving a monument that is worthy of the name, and that will give pleasure and knowledge to many succeeding generations."

The plan called for a 216,000-acre monument that would protect everything of geological, scenic, recreational, and ecological interest around the mountain. To the north, the monument would extend beyond the Green River to the ridge where Clyde Croft and Al Handy had ridden the day before the eruption. To the northwest, it would include the high lakes where the Killians, the Smiths, and other families were scattered that Sunday morning. It included a set of caves created by ancient lava flows south of the mountain, along with broad swathes of both damaged and intact forests to the east and south of the volcano. It would still allow most of the downed timber to be salvaged, but it proposed that thousands of acres of downed trees, representing millions of dollars of potential profits, be left in place for scientific research and for tourism.

Washington State officials, the Forest Service, and logging companies were united in their opposition to the proposal. They contended that the proposal protected far too much forest that could be harvested. It would remove too much area from the tax base of the affected counties, they said, and eliminate jobs. To Susan Saul and her allies in the environmental movement, it seemed as if government officials and the logging companies simply wanted to reestablish the status quo as soon as possible. Clean up the trees, plant the blast zone with grass, and pretend that nothing ever happened.

The state had its own plan for a protected area. A state committee consisting of timber company representatives and members of the Department of Natural Resources, but no member of the conservation community or the scientific community, recommended that an area of less than 50,000 acres be set aside for scientific study. It was insultingly small—less than a quarter of the association's proposal. The entire Green River valley would be open for logging, as would all the other forests surrounding the mountain. The proposal would preserve the volcanic cone and the Spirit Lake Basin, along with a small portion of the Mount Margaret high country, but no more. Only a small portion of the blast zone would be protected—mostly the area where all the vegetation was destroyed—leaving the old growth to be harvested.

The Forest Service's proposed plan was better than the state's, but not much. It called for an "interpretive area" of about 85,000 acres centered on the volcano and Spirit Lake. It designated as much downed timber as possible for salvage. It would have allowed geothermal leasing, mineral prospecting, and open-pit mining operations. As Saul later said of the plan, "Rather than using its administrative resources to devise a management plan of vision or imagination, the Forest Service resorted to a business as usual attitude."

Throughout 1981, all of these proposals remained in play, with none gaining the upper hand. But the status quo was unsustainable. It was up to the politicians to make something happen.

• • •

In her bid for a second term as Washington State governor in 1980, Dixy Lee Ray didn't even make it through the primaries. She had alienated too many people. Voters had come to see her independence as eccentricities. In defeat, she retreated to her Fox Island home. In the years before her death in 1994, she was often seen blasting around nearby Gig Harbor in her red convertible sports car, her dogs sitting in the passenger seats and leaning into the wind.

On the same day that Ronald Reagan was elected president, Washington State elected a moderate Republican named John Spellman as governor. For the previous decade, Spellman had been chief executive of the county where Seattle is located, and the experience made him an expert at forging compromises to get things done. A committed environmentalist, Spellman had defied businessmen, legislators, and his own party to block the construction of an oil pipeline from Puget Sound over the Cascades and Rockies to the Midwest. He had a fishing pole in the trunk of his car and an up-to-date fishing license in his wallet, though neither got used much.

Six months after taking office, Spellman submitted his own proposal of what to do with Mount St. Helens. It cleverly combined elements of the existing proposals into a neat package. It gave up much of the forests south and east of the volcano, but it protected the old-growth low-elevation forest of the Green River valley. It excluded the high lakes northwest of the volcano, but it included the set of caves on the south side of the volcano—formed in previous eruptions by rivers of lava flowing through gradually cooling rock tunnels—even though they were surrounded by saleable timber. The proposal was opposed by most of the state agencies Spellman headed, which agreed with the Forest Service that the amount of uncut timber should be minimized. But Spellman was the type of politician who knew that everyone would be at least somewhat unhappy with a good compromise.

Now the initiative had to move to the federal level, since only the federal government was in a position to bring all the competing interests together. That meant getting Congressman Don Bonker involved, since he was the representative for the congressional district that included much of the Gifford Pinchot National Forest. Bonker was a young, ambitious, and telegenic politician with an interest in international trade and environmental protections. There was just one problem: Bonker thought the Forest Service could manage the area around Mount St. Helens just fine on its own.

Conservationists were completely opposed to that option. If the area around the volcano were managed by the Forest Service as just another part of the national forest, federal managers could build roads, log, and serve the timber industry just as they had in the past. Without protective legislation, the association's goals were unattainable.

As the various proposals vied for attention, Saul and her associates organized a media tour of the Green River valley to show how salvage logging was already occurring in the area proposed for a monument. But access to the area was closed, so Saul asked Bonker for help. When he called to tell her that he'd received permission for them to go, Saul asked if he wanted to come along. Sure, he said. It would help him figure out what to do with the area. But he wanted representatives of the Forest Service to be there as well.

The next day, the bus carrying Saul, the conservationists, and the reporters stopped at Spiffy's restaurant, where Highway 12 hits Interstate 5, to pick up Bonker. When he got on the bus, Bonker looked around and said, "Where's the Forest Service?"

"They're not coming," said Saul.

"That can't be true," said Bonker. "They must be meeting us at the ranger station in Randle."

But when they stopped at the ranger station in Randle, no one was there. On such small incidents can the success of a campaign turn. Bonker realized that the Forest Service would never be capable of managing the area around Mount St. Helens without being told what to do. It was time to get some legislation passed.

THE LEGISLATION

BY THE FOLLOWING SPRING, THREE SEPARATE BILLS WERE IN PLAY to protect the area around Mount St. Helens. One, introduced by Bonker, called for a Mount St. Helens Volcanic Area of 110,000 acres; this was essentially the compromise hammered out by Spellman. The other two would create the small monument specified by the Forest Service or the large monument proposed by the Mount St. Helens Protective Association.

In hearings on the proposed bills that spring, a remarkably diverse array of interests presented their views. Weyerhaeuser's group vice president for western operations said that the loss of timber in the compromise proposal "would have serious implications for forest industry jobs and raw material supplied for the region," though he also expressed Weyerhaeuser's willingness to trade areas it owned within the Forest Service plan for timbered land elsewhere. The commissioner of Skamania County, where the volcano is located, said that the Mount St. Helens Protective Association's proposal was "nothing short of ridiculous." A representative of a plywood company pointed out that the unemployment rate in Skamania County was 33 percent. "We're basically opposed to all the bills," he said. The legal counsel for the Oregon Four Wheel Drive Club said that the eruption of Mount St. Helens was too insignificant to deserve monument status, which would bar off-terrain vehicles from the area. Compared with previous volcanic eruptions in the west, he said, "Mount St. Helens' eruption was only a little pimple on the bunghole of creation."

In contrast, the scientists and conservationists who testified argued for a large monument. A professor of zoology from Ohio State University said, "We have no site on this earth, apart from Mount St. Helens, where we can test some of these ideas of ours. We have needed beautifully clear, bare sterile land, and from this eloquent explosion, which no government on earth is wealthy enough to achieve for us, we have it virtually for free." Geophysicist Steve Malone expressed his fear "that the temptation—in fact, even the desire—to help Mother Nature out, to fix the broken mountain, would be almost irresistible. Influencing what is naturally occurring is anathema to good science that is trying to figure out what would normally occur." A member of the Mount St. Helens Hiking Club who had been on the Green River hike the week before the eruption said, "Eight days before the awful gigantic blast, we hiked for the last time under the towering trees, along the tumbling, aptly named Green River. It would be like coming home to hike that trail again."

The most affecting testimony came from Lu Moore, who described her family's experiences on the Green River:

> My family are third-generation residents of the Toutle–Castle Rock area. Granddad went to work in the woods at thirteen, and my father has logged for Weyerhaeuser for nearly thirty-five years. While most of the people in the valley hold jobs related to the timber industry, most of us do take the time to enjoy the beauty of our surroundings. Mount St. Helens and her foothills provided my family with our prime source of recreation, playing in the snow in the winter, hiking, camping, and fishing in the spring and summer, and hunting in the fall.
>
> Over the years, those of us in the valley witnessed the loss of the best lowland hiking and hunting spots to clearcuts and roading. As the clearcuts crept toward Mount St. Helens, we were

forced to find alternative hunting ground. This was accepted as the price one paid for living in a logging community.

> Relatives and friends from out of state would come to visit and to see the mountain. We were proud to show them our area, but by the late 1960s and early 1970s it became embarrassingly necessary to explain to them why anyone was allowed to clearcut on the flanks of the mountain right up to timberline.
>
> I left the area for a short time to go to school, married, and returned to the Toutle Valley to raise a family of my own. We soon found that clearcuts were encroaching on some of my favorite spots, and in an effort to save some of them I was prompted to learn more about the planning of forest management.

What she learned both before and after her family's ordeal had strengthened her convictions, Moore said:

> The area has a special and new significance for all who wished to experience the effects of the eruption. In August 1980, three months after the event, Mike and I went back to our campsite. Standing on the ridge above the valley, I looked out over thirteen miles to the mountain and wondered how anything could have the force to reach into the valley below me. I looked down the valley at the contrast between green, standing, dead, and downed trees and marveled at what stretched before me. I saw that salvage operations between Miners Creek and Shultz Creek had already begun and feared that perhaps few others would have the chance to witness the phenomenon before me.
>
> As people come through the grocery store where I work, they often ask me about my experience and share some of their experiences, along with their views about what the future should hold for Mount St. Helens. These views have evolved over the last two years from "how could she do that to us" or "who wants that damned thing" to a feeling of acceptance and even pride.
>
> Many locals have a desire to preserve an area around Mount St. Helens for economic as well as other reasons. We feel a need

to diversify our economic base so the community does not suffer so drastically from the ups and downs of the logging industry. . . .

My family has been in the Castle Rock–Toutle area for several generations and plans to remain there. I want my children to have the same quality of life that we have had in the past. I feel a national monument would save a piece of the past for the future.

• • •

In July the House of Representatives passed a bill that protected 115,000 acres of land around Mount St. Helens. A few days later, the Senate passed a bill that protected 105,000 acres. During the conference to reconcile the two bills, congressional staff members were constantly on the phone to Susan Saul and to the Sierra Club's Charlie Raines, asking them which acres were essential and which could be sacrificed. In August a final bill emerged calling for a monument of 110,000 acres. The House approved the bill by a vote of 393 to 8. The Senate passed the bill without dissent.

Weyerhaeuser carried through on its promise to cooperate. It and Burlington Northern traded away about 32,000 acres of land inside the boundaries of the monument in return for 7,400 acres of land outside the protected area. The old-growth forests of the Green River were saved from logging, as were the forests surrounding the lava caves to the south of the volcano. The monument would be under the jurisdiction of the Forest Service, but it would be managed as a separate unit and have its own planners and supervisor. Fishing and hunting were allowed in parts of the monument, but road building was discouraged. Most unusually, the legislation specified that the Forest Service "shall manage the Monument to protect the geologic, ecologic, and cultural resources, in accordance with the provisions of this Act, allowing geologic forces and ecological succession to continue substantially unimpeded."

The bill had made it through Congress, but now it faced a final and perhaps insurmountable hurdle. President Reagan had not forgotten

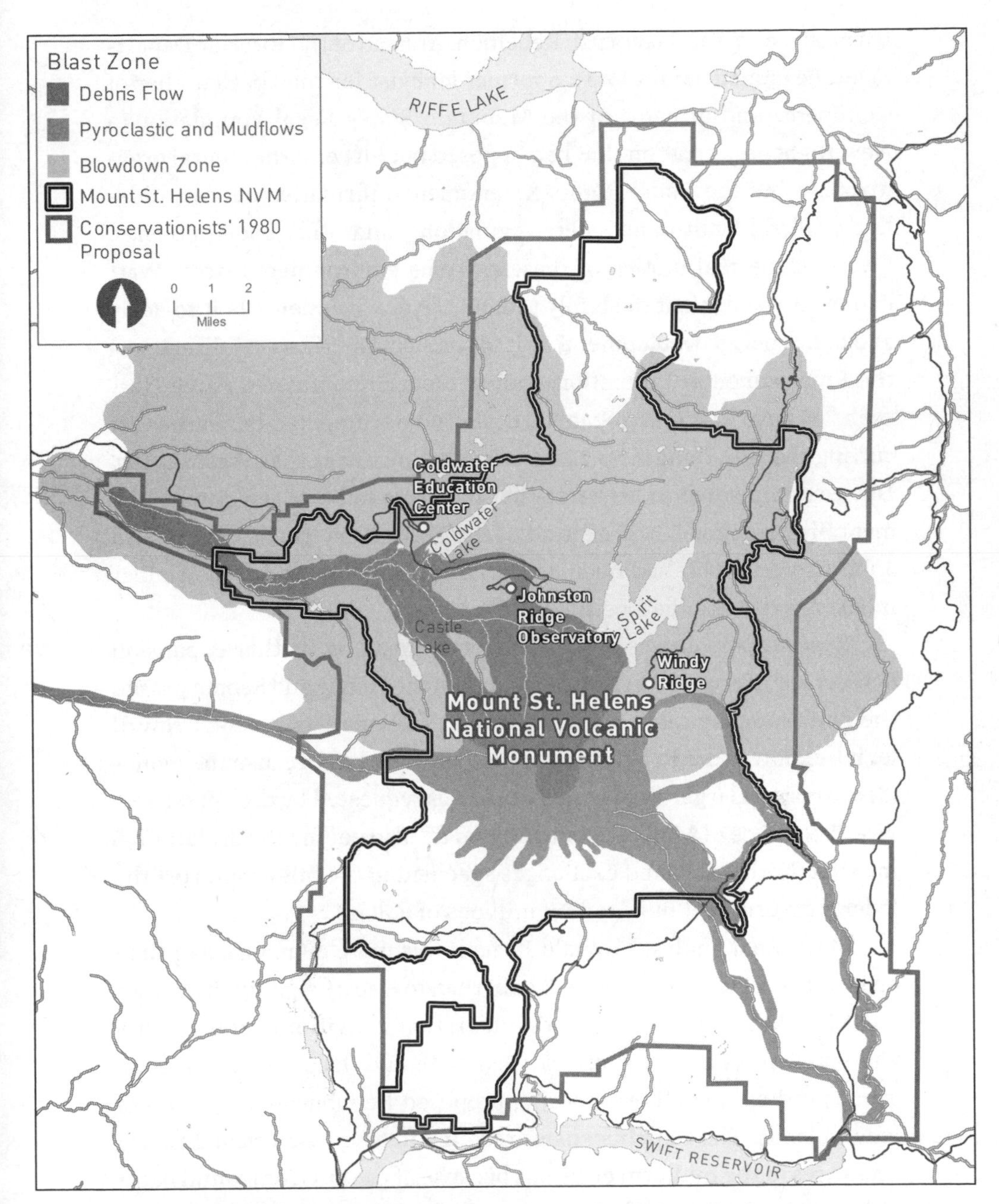

Map of the monument proposed by the Mount St. Helens Protective Association (single line) and the final boundaries of the Mount St. Helens National Volcanic Monument (double lines)

his promises to the Sagebrush Rebellion. As head of the Interior Department, he named James Watt, a former lobbyist for the US Chamber of Commerce and director of the Mountain States Legal Foundation, a New Right organization that had opposed not just environmental organizations but the Equal Rights Amendment, affirmative-action guidelines, federal health and safety regulations, and other social policies. "What is the real motive of these extreme environmentalists?" Watt had once asked. "Is it to simply protect the environment? Is it to delay and deny energy development? Is it to weaken America?" As director of the Environmental Protection Agency, Reagan nominated Anne Gorsuch, a former Wyoming State legislator who trumpeted her success at cutting the EPA budget by 22 percent and once said that Washington, DC, was "too small to be a state but too large to be an asylum for the mentally deranged." And as head of the Forest Service, Reagan picked John Crowell, who had been an attorney and lobbyist for the timber industry in the Pacific Northwest.

Reagan's appointees were dead-set against any further expansion of federal power over western lands. When the House and Senate passed their bills for a Mount St. Helens National Volcanic Monument, Crowell let it be known that Reagan would veto the final bill because the monument area was larger than the 85,000 acres requested by the Forest Service. The Office of Management and Budget also got involved, claiming, incorrectly, that the land exchanges specified in the bill would cost the federal government hundreds of millions of dollars.

In the final analysis, Ronald Reagan was more of a politician than an ideologue. He had far more to gain than to lose by signing the monument legislation. In 1982, his administration's environmental policies were under widespread attack. National and local environmental groups had prospered by opposing proposed Republican giveaways of federal lands. The Republicans were going to lose senators and representatives in the midterm elections because of the environmental backlash. Given that support for the monument legislation in both the House and the Senate was virtually unanimous, a veto could be overridden anyway. Reagan couldn't afford to reject the monument legislation just to make a point.

On August 26, 1982, with little fanfare, Reagan signed the bill. It was the first, but hardly the last, public lands legislation adopted in the Reagan administration. By the end of his presidency, Reagan had added 38 million acres to various categories of permanent protection and nearly 5,000 miles to the National Wild and Scenic Rivers program and the national estuarine reserves system. "Preservation of our environment is not a partisan challenge," he said in 1984. "It's common sense."

A few weeks after the signing, Congressman Bonker held a party in his congressional offices to celebrate the monument's creation. Representatives from all of the relevant groups were there: environmentalists, timber companies, railroads, chambers of commerce. For more than a hundred years, ever since the land around Mount St. Helens was divided into square-mile sections and distributed among eager potential landowners, each group had pursued its own interests in the dense forests, open meadows, and rocky slopes surrounding the mountain. Now they had agreed on at least one thing: the land around Mount St. Helens should be saved for future generations.

PART 7

DECLINE AND RENEWAL

The Pumice Plain to the north of Mount St. Helens in 2005

THE LOGGERS

AFTER THE ERUPTION OF MOUNT ST. HELENS, GEORGE WEYERhaeuser dropped from public view. No news accounts record where he was at the time of the eruption or what he thought about it. The company had escaped disaster when the volcano erupted on a weekend rather than a weekday. But people had died on Weyerhaeuser land, and the company had been involved in decisions about whether to close that land. Anything Weyerhaeuser said could end up hurting him.

Within days, dark rumors began to circulate that Weyerhaeuser had cut a deal with Dixy Lee Ray to keep the company's land open for logging. What Ray would get in return for her cooperation, other than vague promises of political support, was never clear. Nevertheless, the rumors, combined with the anguish of the victims' families, led to legal action. In 1981 relatives of the victims, eventually including the Killian, Pluard, Blackburn, and Crall families, filed a wrongful-death suit against the state of Washington and, later, against Weyerhaeuser. They claimed that the state was negligent in establishing the restricted zones so close to the mountain. When Weyerhaeuser was added to the suit, the company and state were accused of failing to inform the public of the obvious danger posed by the volcano. As the suit put it, "from April 1980, until May 18, 1980, Weyerhaeuser officials continually advised crew supervisors, and often advised other employees and independent contractors, that employees working near the mountain were safe. These representations were clearly false. Weyerhaeuser acted fraudulently and/or recklessly and/or negligently in assuring its employees and independent contractors working near the mountain that they were not in danger."

After extensive fact-finding and legal maneuvering, the case went to trial in 1985. By then the state was no longer a defendant. A superior court had ruled that Dixy Lee Ray had made a legitimate policy decision based on the information available to her, which meant that the state could not be sued. Now the burden was entirely on Weyerhaeuser to defend its actions.

For four weeks a twelve-member jury heard the evidence. The plaintiffs' representative, a lawyer from Houston named Ron Franklin, argued that Weyerhaeuser cared more for its trees than for its employees. "Weyerhaeuser did not vary the work location of one living person as a result of the volcano's activity," Franklin told the jury. "The eruption was an act of God. But the deaths were an act of man, when man ignored weeks and weeks of God's warnings."

Weyerhaeuser's attorney, a Seattle lawyer named Mark Clark, argued that the company did everything it could, given the warnings it received. Weyerhaeuser officials testified that they never heard about the possibility of a lateral blast. The greatest threat they expected was valley flooding, which is why they had shifted their men to higher ground. The geologists called to testify said that the magnitude of the eruption could not have been foreseen. If the volcano had erupted straight up, they said, rather than to the side, few or no people would have died. But Rocky Crandell, who had drawn the original map for the Forest Service of potential dangers around the mountain, also pointed out that the restricted zones had ignored the dangers he identified on the northwest side of the mountain. He and Donal Mullineaux had shown that the mountain had produced ash deposits three feet thick up to twenty miles north of the mountain. "They chose to ignore that," Crandell had said. "I'm not sure why."

In the end, the jury couldn't decide. The majority of the jurors thought that Weyerhaeuser was not to blame, contending that the company could not have foreseen a lateral blast. But several disagreed. They insisted that Weyerhaeuser had a responsibility to move its men away from a dangerous volcano and keep the public off its land.

When the trial ended inconclusively, some of the plaintiffs demanded a retrial. But Franklin, who ended up waiving his own fees, pointed out that if they lost in a new trial they could be on the hook for legal costs. Finally the families agreed to settle for a few thousand dollars apiece. "I'm glad it's over," said John Killian's mother afterward. "We had a chance to tell our story."

Dixy Lee Ray and George Weyerhaeuser certainly talked between the time of the first earthquake and the May 18 eruption. Weyerhaeuser was a member of Ray's economic advisory council, and the group met often with the governor. In an affidavit prepared for the trials, Ray said, "Before May 18 . . . I spoke several times with George Weyerhaeuser. Weyerhaeuser Company was closely following the developments on the mountain because of its ongoing logging activities in the area." In his own affidavit for the trial, Weyerhaeuser said that he "expressed to [Ray] my concerns about the state's procedures for granting access and restricting access to the restricted zones around Mount St. Helens, and the impact of these procedures on Weyerhaeuser's logging operations."

However, both Ray and Weyerhaeuser swore under oath that they did not have any kind of arrangement. As Weyerhaeuser said in his affidavit, "there never was any agreement or 'deal' of any kind between the Governor and me concerning an exclusion of any Weyerhaeuser lands from these restricted zones." Ray said the same thing in her affidavit: "There was no such agreement or deal, explicit or implicit, between me and/or anyone, including George Weyerhaeuser and/or representatives of the Weyerhaeuser Company. No one from the Weyerhaeuser Company, including its president George Weyerhaeuser, asked me to set restricted lands to be excluded from the restricted zones." No evidence turned up before the trial demonstrating that the two of them had any kind of deal, and no evidence has appeared since then.

The truth appears to be more mundane—and more telling. Ray did not need to talk with Weyerhaeuser about the restricted zones. Any understanding between them could remain unspoken. No one in a position of authority would have defied Weyerhaeuser's wishes. The company's money, influence, and prestige were too great; the forces of private property and capitalism, too strong. Weyerhaeuser did relatively little to protect its workers beyond moving some of its crews to high ground and preparing poorly disseminated evacuation plans. But Weyerhaeuser was a major employer and business in Washington State, and the state was unwilling to tell such a company what to do.

Besides, logging was a dangerous business. If the company's loggers had to work near a threatening volcano, their jobs would just be that much riskier. Loggers were accustomed to taking chances. Many would have howled if they were barred from worksites and lost money because of what geologists said might happen.

• • •

As George Weyerhaeuser had foreseen before the eruption of Mount St. Helens, the 1980s marked the end of large-scale old-growth logging in Washington State. Once Weyerhaeuser had salvaged the big trees knocked down by the volcano, it had little old growth left. Weyerhaeuser and other logging companies shuttered their sawmills that could handle big timber or reconfigured them for much smaller second- and third-growth trees. The age of heroic, highball, larger-than-life logging in the United States was over.

As Weyerhaeuser shifted its attention toward the trees that loggers disparagingly call "dog hair," the amount of timber the company harvested from around Mount St. Helens plummeted. Industrial timber harvests in the county that contains Weyerhaeuser's Mount St. Helens Tree Farm fell by more than half during the 1980s and continued to decline thereafter. Weyerhaeuser ripped out its rail lines and relied entirely on trucks to get the remaining timber out of the woods. Auction yards filled up with so many yarders and grapplers that they looked like abandoned logging camps. Today, about one-sixth as much timber comes out of the area as during the boom years of the 1970s.

As harvests fell, Weyerhaeuser began to shed employees. Soon after the salvage operation ended, the company declared a major reduction in force. Hundreds of cutters, choker setters, truck drivers, and millworkers lost their jobs. A few years later, the company said it would have to reduce wages and benefits by at least 25 percent to remain competitive with cheaper lumber from Canada and the southern United States. Speaking at a theater in Longview, George Weyerhaeuser said that the company needed sacrifices from the union if it was to stay afloat. As Weyerhaeuser's chief financial officer said, "We're not a philanthropic enterprise. We're in business to make a profit."

In 1986, negotiations over a new contract deadlocked and the union authorized a strike. Weyerhaeuser refused to back down. It said that if the union did not approve the offered contract, the company would eliminate more than three thousand jobs in the Northwest and replace the fired workers with contractors. The loggers buckled. Though they had voted down the original contract by a four-to-one margin, after Weyerhaeuser's ultimatum they approved virtually the same contract in a two-to-one vote. Union employees lost 20 percent of their previous income.

To make up for some of their lost income, Weyerhaeuser instituted a competitive logging program designed to increase productivity. Logging crews could receive a bonus if they finished a job ahead of schedule or produced more profit than company analysts had predicted. But to get the timber out quickly enough to earn those bonuses, the men had to work fast, which inevitably led some of them to cut corners on safety, although Weyerhaeuser maintained that safety was not compromised. Plus, only some crews got the bonuses, demoralizing the others.

At the same time, technology was reducing the number of jobs in the woods and in sawmills. A single operator in an air-conditioned cab, using a machine known as a feller-buncher, could lay down trees faster than a crew of men with chainsaws. Other machines known as processors could grasp entire trees, limb and buck them to length, and sort and stack the resulting logs. In the mills, a single person at a computerized console could direct laser-guided saws to extract the maximum amount of lumber from a tree of any size, replacing all the line workers on manually run saws.

These employment downturns at Weyerhaeuser were mirrored throughout the industry. In the early 1980s, employment in the Washington State lumber and wood products industry dropped by one-third, representing almost 20,000 people thrown out of work. Since then it has fallen by another third. In the 1970s, most everyone who had lived in Washington State for long was related to or knew a logger. Now most everyone who has lived in the state for long knows someone who used to be a logger. The members of the extended Killian family are still scattered around Vader, Toledo, Castle Rock, and other small towns near Mount St. Helens, but they are state patrolmen, school administrators, government officials; few work in the wood products industry anymore. Choker setters, if they can get a job in the woods, make half of what John Killian was making in 1980. Vader's population is down to 600 and continues to drop.

Later in the 1980s, the wood-products industry in the Pacific Northwest suffered another blow. An obscure Massachusetts environmental group called GreenWorld petitioned the US Fish and Wildlife Agency to place the northern spotted owl on the federal government's list of endangered species. Mainstream environmental groups had cautioned against such a move. They considered the Endangered Species Act too flimsy an instrument with which to challenge the logging industry. But the listing of the owl turned out to be successful beyond GreenWorld's wildest dreams. By the early 1990s, lawsuits based on preserving the owl's habitat had all but halted logging in many parts of the national forests. Though the trend toward less logging and fewer jobs was already well under way, the owl wars hastened the decline.

• • •

As timber harvests dropped, the Weyerhaeuser Company also began to change. In the Northwest, Weyerhaeuser had always been seen as a dowdy, family-oriented company—"a school for the young and a home for the old," in the words of one logger. It was run by men who cared about their employees, who treated them with the tough love a stern father might impose on a son. Now the company was becoming a corporation. To please Wall Street, it focused on the bottom line, not on its employees. Whenever possible it shifted work from well-paid union workers to scrappy, underpaid contractors who had to supply their own equipment. The loss of union jobs meant that loggers no longer had medical coverage, a pension, paid vacations, or a chance of promotion. They were on their own. Some did well, creating strong companies that provided employment for other former union members. But many ex-loggers struggled, losing income and the dignity that comes with having a solid job in a close-knit community. In 1980, Weyerhaeuser had 48,000 employees, 28,000 of whom were unionized. By 2014, its US workforce was down to about 11,000, fewer than 4,000 of whom were union workers.

The company began to close its properties to the public. It said that vandalism, dumping, and the proliferation of meth labs in the back-country made it too expensive to keep the land open; more recently, it has begun charging hunters hundreds of dollars for access to the land. Gradually, the lakes, streams, and woods where local people had camped, fished, and hunted for generations were sealed up behind locked gates. Today Road 2500 up the Green River is closed and rarely used. Farther upriver, in the part of the monument where the Moores were camping on the day of the eruption, relatively few people come all the way out to Ryan Lake and hike downriver to see the trees. The trail though the old-growth forest of the Green River valley is starting to look forlorn.

The leadership of the Weyerhaeuser Company also began to change in the 1980s. George Weyerhaeuser's firstborn son, George Jr., had gone to Yale, as had his father, grandfather, and great-grandfather. But George Jr. majored in mathematics and philosophy, not in the hard-nosed business subjects favored by his forebears. Friendly, intellectual, and quirky, George devoted himself to the company and gradually worked his way up the ranks, becoming senior vice president of technology. But he was the kind of guy who, in high school, read Plato to his younger brother at red lights while driving him to school. Everyone in the company knew he had no chance of ascending to the helm. In 1988 a twenty-year veteran of the company, Jack Creighton, became president. George Weyerhaeuser remained as chairman of the board of directors until 1999, when, almost exactly 100 years after his great-grandfather bought the land around Mount St. Helens from Jim Hill, he stepped down from the leadership. As of the writing of this book, he was still alive and doing well at the age of eighty-nine.

But in 2013—five years after his own retirement from the company—George Jr., at the age of fifty-nine, died of a heart attack while sailing on Commencement Bay, within view of the hillside where his father was kidnapped on his way home from school, within view of the mansion where his great-grandfather died. It was the end of a dynasty. Though Weyerhaeuser remains a major company in the Pacific Northwest, none of the company's top corporate officers and no one on the board of directors bears the family name. It is unlikely that the company will ever be run by a Weyerhaeuser again.

• • •

John Killian's father, Ralph, was never able to come to terms with what happened to his son. After the eruption, Ralph began to search for John in the tangle of downed trees beneath Fawn Lake. The family found pieces of his rubber raft in the trees, which is why they concluded that John was on the lake fishing when the volcano erupted. But there was no trace of John. The Weyerhaeuser Company eventually gave Ralph two years off with pay to continue the search. He cut away tree limbs, dug through the thick ash, and pried apart piles of rock, but he never found his son.

John's mother said, "You feel like you want to bring him home if you can find him. Is he lying on the ground somewhere? What happened? It's just something we can't leave be."

Gradually Ralph became convinced that John was knocked unconscious by the explosion, lost his memory, and fell into a new life. Whenever he passed a group of loggers, he carefully searched their faces to see if one of them might be John. Once he spotted a logger riding in a Weyerhaeuser company pickup who was a dead ringer for his son, and he spent days trying to track down the truck. "I don't know if it was an illusion," he said.

"I've never believed [John] died," he once told a reporter. "I don't believe it to this day. He's dead, really. But it's hard to accept. It's like he walked outside and should be back."

THE SCIENTISTS

THE SCIENTISTS WHO HAD BEEN WORKING AT MOUNT ST. HELENS were devastated by the eruption. Despite all their efforts at monitoring the volcano, they were unable to provide any warning of its eruption. "We'd failed," said University of Washington seismologist Steve Malone. "For two months we'd counted and located thousands of earthquakes, looked for changes to anticipate an eruption. Then it just happened. It killed many people. It killed David Johnston. We could hardly work."

After the eruption, geologists were quick to defend themselves from charges that they should have foreseen the lateral blast. An event of such size had never occurred before at Mount St. Helens, they observed. They would not have stationed Dave Johnston so close to the volcano if they expected it to explode. Furthermore, they had little experience with erupting Cascade volcanoes, since the last eruption had occurred more than sixty years before and the United States had been investing most of its volcano research money in Hawaii. "Naturally, if I had it all to do over, I'd do it differently," said Rocky Crandell. But "if I had drawn on the hazards map showing a lateral blast off of the north side, I simply could have not supported it with fact. A hazards analysis is based on what has happened in the past at that particular volcano."

Yet the geologists had discussed the possibility of a lateral blast and strong pyroclastic flows among themselves, and their speculations occasionally made it into print. As Crandell and Mullineaux wrote in a 1975 article, "Mount St. Helens has probably included . . . violent eruptions like Vesuvius in AD 79." In their 1978 publication on the volcano, they mentioned a previous "strong laterally directed explosion" on the volcano, though that one was relatively small compared with the 1980 blast. Dave Johnston, among others, recognized the possible parallels of Mount St. Helens to Mount Bezymianny in Russia, though Cold War tensions had limited the information available about the mountain. But Crandell had a box of scientific articles that he shared with others in Vancouver, including a comprehensive article on the 1956 eruption at Bezymianny and other articles that considered lateral blasts. Barry Voight, in his May 1 report, mentioned the possibility of a landslide on Mount St. Helens that would create a horseshoe-shaped crater, though the lateral blast at a Japanese volcano that he used as an example was smaller than the 1980 blast. And an armored personnel carrier was on its way to Coldwater II, though it would not have saved Johnston. Like loggers, volcanologists had to take risks to do their jobs, and they deemed the risks that they were taking to be acceptable given the potential benefits.

Perhaps the greatest failure of the monitoring effort at Mount St. Helens was the insufficient attention devoted to the worst things that could happen. Though lateral blasts were not well understood at the time, geologists knew that volcanoes could explode sideways, but no one systematically explored that possibility and explained it to the public. In their internal discussions and their public communications, the geologists stuck largely to the most likely things that could happen, not the outliers. A large event was possible but unlikely, and scientists still have difficulty dealing with low-probability high-consequence events. But without some knowledge of what could happen, the people around the volcano that Sunday morning were unprepared for what did happen.

Today, geologists would be able to issue better warnings. For one thing, they have improved their procedures for dealing with government officials, the press, and the public when a volcano acts up. In 1980 the US Geological Survey had only five public information officers in the entire nation, none of whom was assigned to Mount St. Helens before the May 18 eruption. As a result, geologists like Mullineaux and Crandell had to serve both as scientists and as press officers. They did their best, but dealing with reporters takes practice and specialized skills. As an official recap of the episode put it, geologists as a group "would have preferred the simple hazards of field work to the awesome battery of microphones, cameras, and poised pencils each day."

After the eruption, the US Geological Survey began to designate an "information scientist" with experience in press relations to convey information to the public. It also created a standardized volcanic activity alert-notification system, which uses the same *advisory*, *watch*, and *warning* labels that the National Weather Service uses for hurricanes and tornadoes (though some have complained that such a simple system cannot capture the diversity of volcanic hazards). It has developed preparedness plans for hazardous volcanoes and regularly conducts exercises with emergency planners to prepare for the worst.

The eruption of Mount St. Helens dramatically increased funding for volcano research in the United States and around the world, and volcanologists have put that funding to good use. First, they took advantage of one of the most remarkable aspects of the entire eruption: the fact that it took place on a clear and sunny spring morning. By all rights, the blast cloud should have descended on its victims from the midst of a rainstorm, or in the dark of night, or on a typically cloudy Northwest day. But because it occurred in the daytime, people all around the mountain were able to take photographs of the ongoing eruption. Geologists have magnified, combined, and cross-correlated these images, calculating speeds, flows, and forces. They know much more about lateral blasts now than they did before the eruption.

In the decades since the eruption, scientists worldwide have developed and applied a fantastic suite of new technologies to monitor volcanoes. Global Positioning System instruments, a technology known as lidar (for light imaging and ranging), and digital photogrammetry can measure ground movements with an accuracy that would have astonished the geologists who worked around Mount St. Helens in 1980.

Helicopters can cable-release portable seismometers and other instruments mounted on tripod-like stands known as "spiders" to keep an eye on the precursors of eruptions. Computers in the field and supercomputers in laboratories process data, analyze signals, and model eruptions, mudflows, ashfalls, and climatic effects. With this monitoring equipment, geologists have been able to predict, a few hours or days in advance, almost all of the smaller eruptions that have occurred since 1980 at Mount St. Helens. Equipment developed after Mount St. Helens's eruption is now installed at especially dangerous volcanoes worldwide to warn nearby communities of danger. Many of the six hundred or so active volcanoes around the world are still inadequately monitored, and disasters will continue to occur when people do not receive or heed warnings or when volcanoes act in unexpected ways. But no one will ever again blithely and knowingly watch the side of a volcano bulge out like a balloon ready to pop and expect to outrun the ensuing catastrophe.

Some of the things geologists have learned are not reassuring. The hummocky appearance of the landslide down the north fork of the Toutle River was one of the first and most shocking lessons. Geologists had seen jumbled hills like that before. For example, a very similar set of hills occurs about twelve miles north of Mount Shasta. After the eruption of Mount St. Helens, geologists realized that those California hills were created in a huge debris avalanche that occurred about 300,000 years ago. Since then, geologists around the world have identified hundreds of other avalanche deposits near volcanoes. Over the eons of geologic time, volcanic collapses are common, geologists have learned. Mount St. Helens has collapsed before and will collapse again. Mount Augustine in Alaska, which Dave Johnston studied, has collapsed a dozen times in the past two millennia. Geologists have found evidence for collapsed volcanoes not just on the surface of the earth but underwater and even on other planets.

Geologists also have learned much more about the mudflows that often accompany volcanic eruptions. For example, they knew before Mount St. Helens's eruption that its northern neighbor, Mount Rainier, had shed prodigious amounts of rock and mud into surrounding valleys, but the eruption highlighted the full extent of the threat. Today, more than a hundred thousand people south of Seattle and Tacoma live on top of mudflows that have occurred within just the past few thousand years. Moreover, some of the mudflows appear to have occurred without any accompanying eruption. Over time, the heat and gases emitted by a volcano can soften the overlying rocks to the point of collapse. If part of Mount Rainier and its mantle of glaciers suddenly gave way, the people in the valleys below would have just a few minutes to escape.

Almost all of the major volcanoes of the Cascade Range have been more active than most geologists had thought, with a significant eruption occurring on average every twenty-five years. If the eruptions of Mount St. Helens since 1980 are treated as a single event, a new eruption in the Cascades is, statistically speaking, overdue. Future eruptions of these volcanoes have the potential to be much larger than Mount St. Helens's 1980 eruption. When Mount Mazama in southern Oregon erupted about 7,700 years ago, well after Native Americans had settled North America, it emitted a hundred times as much ash as Mount St. Helens, leaving deposits a foot thick more than seventy miles away. Today, the sublime Crater Lake that fills the caldera left when the volcano collapsed into its own evacuated magma chamber provides no sense of its violent origins.

Even larger eruptions than that of Mount Mazama are possible. In the United States and its territories, more than 150 volcanoes are capable of erupting, from Alaska and Hawaii through all the western states as far east as Wyoming, and some of these volcanoes could devastate large parts of the country. If the caldera under Yellowstone National Park were to erupt the way it did 640,000 years ago, ash could cover the entire United States, and the sudden cooling of the planet caused by volcanic aerosols would decimate harvests worldwide.

Meanwhile, the threats posed by volcanoes around the world continue to grow, in part because the fertile soils around volcanoes attract human occupants. In 1985 a relatively small eruption at the Nevado del Ruíz volcano in Colombia, which is the northernmost active volcano in the Andes, generated a mudflow that killed more than 22,000 villagers, most in the down-valley town of Armero. The enormous 1991 eruption of Mount Pinatubo in the Philippines required the evacuation of hundreds of thousands of people and the permanent closing of the United States' Clark Air Base. Near-crashes of commercial airliners that accidentally flew into ash plumes have led to strict rules about closing airspaces downwind from erupting volcanoes. Every year an average of fifty to sixty volcanoes erupt around the world, many in remote locations but others in the midst of dense populations. In the past five hundred years, more than twenty eruptions as large or larger than that of Mount St. Helens have occurred worldwide. A better understanding of how volcanoes behave can reduce, but not eliminate, the threats they pose.

• • •

Between the eruption of May 18, 1980, and the end of 1986, about twenty separate eruptions of Mount St. Helens built a new lava dome that rose hundreds of feet above the crater floor. Then the volcano went quiet, though occasional earthquakes beneath the new dome indicated that magma was still on the move.

From 2004 until 2008 the volcano erupted again, extruding smooth-sided lava spines that eventually collapsed into heaps of stone. If the dome keeps periodically expanding at the rate it has been since 1980—and if it does not blow itself to pieces again in another eruption—Mount St. Helens could regain its pre-1980 profile in a century or two.

THE CONSERVATIONISTS

WHEN RONALD REAGAN SIGNED THE BILL CREATING THE MOUNT ST. Helens National Volcanic Monument in 1982, much of the area set aside for posterity was still a wasteland. Within a few miles of the summit on the north side, the blast blew away every living thing and scoured the ground to bedrock. Subsequent eruptions then deposited a thick layer of frothy white pumice on the bare ground. Today the area between the summit and Spirit Lake, though considerably altered by erosion, is still known as the Pumice Plain.

Farther away, the blowdown zone of jackstrawed trees extended as far as seventeen miles from the volcano in a vast, fan-shaped wedge to the northwest, north, and northeast. In a few minutes, the volcano blew down enough timber to construct an entire city. The toppled trees were covered by a thick layer of chalky gray ash, like a shroud thrown across the landscape.

Right on the edge of the blowdown zone, where the blast cloud finally ascended into the sky, temperatures were hot enough to kill the trees but the blast was not strong enough to blow them over. Today, the dead trees that were not salvaged are still there, a thin band of ghostly white snags facing toward the volcano.

The scientists who testified in favor of the monument's creation were right about its potential for research. Studies conducted in the monument since 1982 have upended previous ideas about how landscapes recover from disaster. One surprising observation was how quickly and extravagantly life returned to the mountain—and how much the process was dependent on chance. In part, the recovery was shaped by the timing of the eruption. Because much of the high country was still covered by snow when the volcano erupted, pocket gophers, deer mice, shrews, voles, and other small mammals were still in their burrows, and some of them survived the eruption. Eating buried roots and bulbs, they mixed underlying soil with the sterile ash, which enabled seeds to take root. Frogs, newts, salamanders, and other amphibians survived in the mud of lakes and rivers, emerging to a hostile environment that killed many but not all of them. The wind carried seeds, insects, spiders, and insect eggs from the surrounding forests. In the first several months after the eruption, more than 120 kinds of spiders wafted onto the Pumice Plain, some having traveled more than sixty miles. Most died, but their carcasses provided nutrients that later colonizers could use.

The researchers expected that Mount St. Helens would recover from the outside in. Species would gradually encroach on the dead zone from the margins. Instead, recovery expanded from thousands of focal points within the blast zone. A plant or animal would establish a beachhead inside the devastated areas. These newcomers would create tiny oases of life that attracted additional colonists.

An especially important plant in the early stages of recovery was *Lupinus lepidus*, prairie lupine, a purple- and blue-flowered legume that took root even on the Pumice Plain. Lupine plants have nodules in their roots that contain nitrogen-fixing bacteria, so they do not have to draw nitrogen from the soil and can grow almost anywhere. Soon vast fields of purple flowers appeared on the blasted plains and hillsides. The individual lupine plants around Mount St. Helens generally died within a few years, after which they provided organic material for other plants. Fireweed, pearly everlasting, thistle, and sedges began to take root in the rough soil. Within a few years, conifers began to appear, growing from seeds blown by the wind or carried by animals into the blast zone.

The succession of species around the mountain did not follow a single path. Rather, it diverged from one place to another, and these local differences have been persistent. Even when conifers cloak the hills and valleys again, as they did before the eruption, the understory will probably maintain a large amount of variation that began in the earliest stages of recovery.

Another lesson learned from the recovery process concerns the value of disorder. Typically, land managers try to recover from a natural disaster by cleaning up the debris and laying down some sort of ground cover to stabilize the soil. But that approach failed at Mount St. Helens. Shortly after the eruption, an expensive and controversial aerial reseeding program dropped grass seeds onto steep slopes near the mountain. But the grass was sown during the wrong season, so few of the seeds germinated. Erosion washed away many of the seeds and deposited them along streambeds, causing a huge boom in the local mouse population. After the seeds ran out, the mice began to starve. They gnawed on the recovering trees to survive, suppressing their growth and slowing the recovery of the landscape.

The habitats that have recovered fastest are those that were left in disarray. Where the downed trees were removed from the monument, biological diversity is relatively impoverished. But where the trees were left to rot and revert to soil, seeds could take root and plants and animals could flourish. Many of the species that occupied these devastated areas were new to the region—western meadowlarks, spiders found previously in the deserts of eastern Washington, knapweed, stem-boring beetles. Today the area surrounding Mount St. Helens has much more biological diversity than it did before the eruption. For that reason, ecologists prefer to call the reestablishment of life around the volcano a *renewal* rather than a *recovery.*

• • •

After the creation of the Mount St. Helens National Volcanic Monument, Susan Saul contributed to many other environmental movements. She worked on the statewide campaign that resulted in the Washington Wilderness Act of 1984, helping to create new wilderness areas in the state and expand existing ones. In 1985 she and a group of other environmentalists founded the Gifford Pinchot Task Force, which works with the Gifford Pinchot National Forest to "support the biological diversity and communities of the Northwest through conservation and restoration of forests, rivers, fish, and wildlife." She joined the board of directors of the Washington Trails Association and strengthened the organization's advocacy program to build, maintain, and protect trails in the state.

In 2006 she retired from the Fish and Wildlife Service after a thirty-three-year career with the agency. Since then she has been working to oppose exploratory mining near the monument. She has joined the board of directors of the Friends of Mount Adams to help with conservation issues there, has begun doing rare-plant conservation with the University of Washington's Rare Care program, and has continued to work with the Washington Trails Association, where she often volunteers on trail maintenance crews.

• • •

Today, a visit to the Mount St. Helens National Volcanic Monument—when the sky is clear and the ripped-open mountain lies yellow and gray beneath a hot afternoon sun, like an immense rock sculpture, like an exhibit in a vast outdoor museum—is an unforgettable experience. Following the creation of the monument, the federal government built a new highway into the heart of the blast zone. It follows the old Spirit Lake Highway as far as a giant dam built across the north fork of the Toutle River to hold back the sediment that continues to wash downstream. Above the dam the highway climbs the ridge above the north fork, offering views across the valley and occasionally ahead to the blasted-out volcano. The road enters the blowdown zone; here, Weyerhaeuser has replanted its land with a crop of noble firs, which rise in serried ranks from the streambeds to the ridgelines.

On the ridge where Reid Blackburn stood taking photographs of the volcano, the road runs along the boundary between Weyerhaeuser's land to the north and the monument to the south. Here the monument land has not been replanted, and the views toward the mountain are open and grand. The road passes the Science and Learning Center at Coldwater, where the nonprofit Mount St. Helens Institute conducts educational activities for students, teachers, and members of the public. Then the road dips into a valley where the avalanche dammed Coldwater Creek, creating a new lake that today is ringed by willow and alder. It climbs a final ridge, past the slope where Jim Pluard's crew was logging the week before the eruption. On top of the ridge, it passes within a few feet of where Dave Johnston was standing when the north slope failed and the eruption began. A few hundred yards later it ends at a low-slung, unobtrusive concrete building perched on the southern edge of what, in the geologist's memory, is now called Johnston Ridge. On a concrete viewing platform in front of the Johnston Ridge Observatory, people stand and gaze across the valley at the astonishing spectacle of the volcano. Inside, tourists speaking a hundred different languages view displays about the eruption, watch a thunderous movie that recreates the approach of the blast cloud, and buy T-shirts, coffee mugs, and picture books while nearby Forest Service rangers answer questions about what happened on that long-ago day.

From the Johnston Ridge Observatory, the more adventurous visitors can follow a trail along the edge of the ridge to an overlook of Spirit Lake. When the north slope of the mountain gave way on May 18, the avalanche hit Spirit Lake like a boot stamping on a puddle. The lake water splashed up the far side of the basin, ripping the trees from the slopes. As the water sloshed back into Spirit Lake, it carried the trees with it. Ever since, a huge raft of bleached logs has floated on the surface of the lake. They travel from one side of the lake to the other, depending on the direction the wind is blowing. At times they seem to create an almost solid surface above the cerulean water.

The monument attracts three quarters of a million people annually. They come not only to the Johnston Ridge Observatory but to the Green River valley, to an east-side overlook on Windy Ridge, to the south-side lava caves and snowfields, and to the southern slope of the volcano, which thousands of people climb each year. The Forest Service manages the monument, and despite occasional calls to make the area into a national park, most people are satisfied with the Forest Service's management, despite perennial funding shortfalls. The Forest Service is a different agency than it was during the get-out-the-cut era of the 1950s, '60s, and '70s. As harvests from the national forests fell during the 1980s and '90s, the agency reverted in part to a caretaker organization, as it was in the first half of the twentieth century. Funding levels plummeted, the extravagant ranger stations closed or were downsized, and trails and roads began to wash out and not be repaired. Today the national forests in the Pacific Northwest are wilder than they were at the time of the eruption.

Contrary to the fears expressed by local government officials at the 1982 congressional hearings, the establishment of the monument boosted the local economy rather than harming it. Between 1982 and 2008, the three counties around Mount St. Helens added 30 percent to their population while real per capita income grew by 25 percent. Hunters, mountain bikers, hikers, and commuters who work in Vancouver and Portland have been moving to the area. Though many small towns and rural areas remain economically depressed, new houses dot the landscape.

Despite the monument's popularity, it continues to face difficulties. No new land has been added to the monument since its creation, despite forceful campaigns to do so. When the monument was created, the high lakes northwest of the volcano, where the Killians, Smiths, and others were camping the day of the eruption, were excluded. In 2007, Weyerhaeuser, which owned most of the high lakes region, sold its holdings in the area to two Tacoma-based businessmen, and since then rumors have circulated that the area will be converted into housing lots or a private hunting reserve. Today, the lake where John and Christy Killian died is private property and cannot be visited legally by the public.

An even greater threat is the continued possibility of mining right next to the monument. Prospecting companies have continued to drill boreholes near the headwaters of the Green River, in an area left out of the monument by Congress to protect mining claims. If the drilling were to turn up promising deposits—an unlikely prospect, given the mostly fruitless prospecting in the area for more than a hundred years—a mining company could develop a huge open-air pit next to the monument. Acid would almost surely leach from mine tailings into the Green River and downstream water supplies. The Gifford Pinchot Task Force has been fighting the proposals, but the ultimate outcome remains uncertain.

Even the scientific research for which the monument was established is not guaranteed. Some portions of the monument are closed to public activities so that scientists can carry out their studies. Advocates who want to see more tourists visit the monument urge opening up some of these areas, while conservationists and scientists resist. As with all of the national forests, people might agree on the need to protect an area, but there the agreement ends.

Disputes are inevitable when it comes to the uses of public land. Yet, for all that, the land remains. The creation of a national monument is a farsighted act for an often shortsighted species. When all of us are gone, the monument will still be welcoming people to the ever-changing landscape—at least until the next eruption.

EPILOGUE

VOLCANIC ERUPTIONS ARE JUST ONE OF THE HAZARDS FACING PEOPLE who live in the Pacific Northwest. Since 1980, geologists have learned that the tectonic processes occurring off the coast of Oregon, Washington, and British Columbia are much more violent than they thought. The subduction of oceanic crust beneath the North American plate regularly generates huge earthquakes like the ones that devastated the Indian Ocean region in 2004 and Japan in 2011. These earthquakes have an average recurrence interval of about five hundred years, and the last one occurred on January 26, 1700, triggering a tsunami that destroyed villages all the way across the Pacific, in Japan. That earthquake was sixty times more powerful than the 1906 earthquake that reduced San Francisco to rubble. Computer simulations of a large earthquake beneath Seattle put the death toll at 1,600, with 24,000 more people injured and nearly 10,000 buildings destroyed. The tsunami likely to be generated by a subduction zone earthquake would kill hundreds or thousands. When that earthquake hits, it will eclipse the eruption of Mount St. Helens as the most powerful natural disaster in U.S. history.

People elsewhere may congratulate themselves for living in less dangerous parts of the country, but such complacency would be misguided. According to a 2006 study, 91 percent of Americans live in places with a moderate to high risk of earthquakes, volcanoes, tornadoes, wildfires, hurricanes, flooding, high-wind damage, or terrorism. Meanwhile, everyone on earth faces the certainty of higher temperatures, more intense storms, degraded ecosystems, and higher sea levels as we continue to pump more carbon dioxide into the atmosphere. In many ways, we are all like the people camping northwest of Mount St. Helens in the weeks and days before the volcano's eruption, blissfully unaware of the risks we face.

Emergency planners have gotten better since 1980 at preparing communities in the Northwest for disaster. Pole-mounted sirens in the floodplains beneath Mount Rainier and along the coast can warn people of mudflows or tsunamis. Signs on beachside roads direct evacuees toward higher ground. People here know that they should stockpile food and water in case an earthquake cuts off supplies.

But much more needs to be done. Old masonry buildings need to be reinforced so they won't collapse on their occupants. Structures need to be built in coastal areas so that people can get above oncoming tsunamis. Annual preparedness drills—maybe on May 18—could help cities, towns, and rural areas prepare for the inevitable megaquake. Residents of the Northwest need to replace complacency and fatalism with a well-informed respect for the hazards that lurk beneath our feet.

From the southern rim of Mount St. Helens' new crater, the entire tableau is laid out for the breathless climber. There is Spirit Lake with its ghostly log raft. There is the ridge where Dave Johnston was standing, ridiculously close to the torn-open volcano. There is the jagged and steaming lava dome, incongruously encircled by a rock-covered glacier. In retrospect, the story seems preordained, as if the people around the mountain on May 18 were playing out designated roles.

But that's a misconception, the product of retrospective fatalism. Things could have gone differently in 1980, just as unexpected events have continued to occur since then. Washington State, for example, is a much different place than it was in 1980. It is more prosperous, with a diversified economy that can weather economic downturns much better than the boom-and-bust businesses of logging, farming, and fishing. Much more land has been protected from development, through proposals to set aside additional land inevitably generates controversy.

Many more people get outside to enjoy the natural beauty of the region, so popular trails are more crowded now than they were when I was a kid. Washington State has become simultaneously more high-tech and more outdoors-oriented—a good combination, given the way things are going. Perhaps that combination of forward-thinking and environmental awareness can help us prepare for the next disaster.

The world is impermanent—the eruption of Mount St. Helens showed how quickly and drastically things can change. Yet we still can be good stewards of the things we love.

ACKNOWLEDGMENTS

I TALKED WITH HUNDREDS OF PEOPLE ABOUT THE ERUPTION OF Mount St. Helens while writing this book. I'd especially like to thank Jim Adams, Joe Alper, Dave Anderson, Ivan Bachman, Elna Baine, LeRoy Baine, Al Bates, Don Bonker, Jim Byrne, Christine Colasurdo, Brad Cook, Corrin Crawford, Todd Cullings, George Draffan, Carolyn Driedger, Gregg Easterbrook, Roland Emetaz, David Folweiler, Dick Ford, Ron Franklin, Peter Frenzen, Wendy Friedland, Jessica Friedman, Donna Gerardi Riordon, Liz Hicker, Dave Holland, Sally James, John Hudson, Anita Keeney, John Kendall, Ellie Lathrop, Rick LaValla, Jim LeMonds, Jean Macfarlane, Ross Macfarlane, Ryan Malarkey, Steve Malone, John Martin, Rick McClure, Noel McRae, Robert Melbo, Joe Melton, Charlene Merzoian, Dave Merzoian, Mary Meyer, Richard Meyer, Tom Mulder, Paul Nordstrand, Linda Noson, Dave Olson, Diane Olson, Frank Olson, Ian Olson, Lisa Olson, Lynette Olson, Rick Olson, Roberta O'Neill, Ed Osmond, Dan Para, Nancy Parkes, Charlie Raines, Sue Richman, Michael Riordan, Lisa Romano, Keith Ronnholm, Bill Ruckelshaus, Eric Rutkow, Dexter Salsman, Susan Saul, Grace Schmidt, Roger Sedjo, Joni Sensel, Adam Shapiro, Mitch Sheldon, Arlen Sheldrake, Kevin Snider, Heather Tallis, Susan Tanabe, Amy Tanska, Tod Thayer, Mike Town, Dale Vanlaanen, Charley Vermilyea, Umberto Vizcaino, Barry Voight, Richard Waitt, Luke Wakefield, Liz Westby, David Williams, Andy Wilson, Melissa Young, and Ray Yurkewycz. I'm particularly grateful to the friends, relatives, and coworkers of John and Christy Killian who shared their stories of the Killians with me.

The thousands of pages of court documents kept on microfilm at the King County Courthouse were an invaluable source of information in writing this book. They put to rest rumors that have circulated in Washington State ever since the mountain exploded. I greatly appreciate the county's work in preserving those records and making them available to me.

Librarians and archivists are some of the most helpful, well-informed, and underappreciated people on earth, and I benefited greatly from their assistance. Thanks to Glenda Pearson, Chris Blomquist, Blynnc Olivicri, and Betsy Wilson at the University of Washington Library; Lupita Lopez at the Washington State Archives; Megan Moholt at the Weyerhaeuser Archives; Ken House at the Seattle National Archives; Donna Hill at the Augustana College Library; Emily Tobin at the Rock Island Public Library; Marie Marquardt at the Alma Public Library; Hamp Smith and Sarah Quimby at the Minnesota Historical Society Library; Chris Skaugset and Karen Straube at the Longview Public Library; Mario Milosevic and Kelley Davis at the Stevenson Community Library; Vicki Selander at the Castle Rock Library; Susan Tissot at the Clark County Historical Museum; and Janice Goldblum at the National Academy of Sciences.

I had three superb research assistants while writing this book: Sarah Olson, Eric Olson, and Amelia Apfel. Working with my son and daughter on this book was a special pleasure for me.

Matt Stevenson at CoreGIS drew the wonderful maps in this book, and Charlie Raines helped me get the maps and other geographical details straight. Thanks to Jack Shafer for sending me the link to the Mount St. Helens volcanocam when he lived in this Washington and I lived in the other one.

I'd like to express my particular gratitude to the people who read part or all of the book in manuscript and helped me revise it: Christine Colasurdo, Jim LeMonds, Steve Malone, Barry Voight, and the members of my mom's reading group: Ruthie Cravens, Anne Davis, Sherry Sheffield Dougherty, Carroll Ray Kramer, Leila R. Kramer, Judy Kuehn, Kae Paterson, and Diane Sorrells.

My agent, Rafe Sagalyn, to whom I've dedicated this book, has been a huge influence on my career as a writer. None of the books I've written would have turned out the way they did without Rafe's advice, expertise, and vision.

From the moment I met my editor, Alane Mason, in her offices overlooking the New York Public Library, I knew that she and I would make a great team. Her sensitive and sensible editing greatly improved the initial draft of this book. Marie Pantojan kept the publication process rolling. Rachelle Mandik did a spectacularly careful copyedit of the text. Will Scarlett skillfully handled publicity.

Finally, I'd like to thank my wife, Lynn Olson, for her love and encouragement. A lifelong easterner, she instigated our move to the West that resulted in this book. She happily accompanied me on many trips to the mountain and listened for hours while I described what happened here or why this detail is important. And throughout the years of working on this book she never had anything less than complete confidence that what I was doing was worthwhile, which is about all a writer really needs.

Seattle, May 2015

SOURCE NOTES

With a few exceptions, full citations for books, scholarly journal articles, and government reports appear in the bibliography on page 265, while full bibliographic information for newspaper articles, websites, magazine articles, court records, annual reports, videos, and other sources of information is provided in the following source notes. Quotations in the text without a source note are from conversations.

PROLOGUE

ix *Earthquakes were shaking the volcano*: *Daily News and Journal-American*, 20.

x *In a 1976 magazine article*: Tom Wolfe, "The 'Me' Decade and the Third Great Awakening," *New York Magazine*, August 23, 1976, 29.

x *"Will the last person"*: Luis, 68.

xi *Though hurricanes and tornadoes*: Harris, 74.

xi *The last earthquake*: US Geological Survey, "Deaths from U.S. Earthquakes." Annual updates are available at http://earthquake.usgs.gov/earthquakes/states/us_deaths.php.

xi *It was the single most powerful natural disaster*: Jeff Masters, "Hurricane Sandy's Huge Size: Freak of Nature or Climate Change?" Available at http://www.wunderground.com/blog/JeffMasters/archive.html?year=2012&month=11.

xi *one of the largest landslides*: Voight et al., 347.

xii *On August 24 in the year 79 CE*: Scarth, 56.

xiii *"It was not the darkness"*: Ibid., 79–80.

xiv *Throughout the nineteenth and twentieth centuries*: Ficken, *The Forested Land*, xiv.

xiv *"The forests of America"*: John Muir, "The American Forests," *Atlantic Monthly*, August 1997, 145.

xv *William Boeing dropped out of Yale*: Verhovek, 25.

xv *Before 1980, incomes*: Noah, 25.

xv *Union membership peaked in 1979*: Barry T. Hirsch and David A. Macpherson. 2003. "Union Membership and Coverage Database from the Current Population Survey." *Industrial and Labor Relations Review* 56 (2): 352.

xv *The year 1980 saw the highest divorce rate*: *The Historical Statistics of the United States*. Millennial Edition, vol. 1 (New York: Cambridge University Press, 2006), 689.

xv *The release of carbon dioxide*: The Earth System Research Laboratory of the National Oceanic and Atmospheric Administration provides information on carbon dioxide and other greenhouse gases. See http://www.esrl.noaa.gov.

xv *In 1979, Bill Gates*: Manes and Andrews, 37–49.

xv *In 1982, Howard Schultz*: Schultz, 90.

xv *In 1983, Costco*: The corporate history is available at www.costco.com/about.html.

xvi *A few years after that, Jeff Bezos*: Stone, 29.

xvi *Today, those four companies*: Starbucks and Paccar, which manufactures trucks in the Seattle area, have about the same revenues, with Nordstrom, Weyerhaeuser, and Expeditors International of Washington the only other Washington State companies on the Fortune 500. Boeing, to the state's chagrin, is now headquartered in Chicago.

xvi *"Sooner or later"*: Jerry Adler, "Seattle Reigns," *Newsweek*, May 20, 1996.

xvii *The* Limits to Growth *report*: Meadows et al.

xvii *In about 1980 the first articles*: Steve Olson, "Computing Climate," *Science 82*, May 1982, 52–60.

xvii *the nation went through a brief frenzy*: Ehrlich et al.

PART 1: THE LAND

3 *At 3:47 in the afternoon*: Endo et al., 93.

4 *Then she entered the results*: Tom Griffin, "Gates' Way," *Columns: The University of Washington Alumni Magazine*, September 2004, 27.

4 *He looked more like a lumberjack*: Michael Lienau, *The Fire Below Us, Remembering Mount St. Helens, A Dramatic Documentary*, Global Net Productions, 1995.

4 *"This is an extremely dangerous place"*: Ota, Snell, and Zaitz, 8.

4 *Dave Johnston was an unusual choice*: Allan Brettman, "This Is It," *Longview Daily News*, May 27, 1995.

6 *"There is no question"*: Thompson, 38.

6 *A blackened crater*: Christiansen and Peterson, 18.

7 *hundreds of loggers*: Jones, 102.

7 *"The mountain is blowing"*: Don Bundy, "Eruptions Small Deal to Loggers," *Portland Oregonian*, March 31, 1980.

7 *"There's no concern"*: Lew Pumphrey, "Weyco Crews Evacuate, Return to Work," *Longview Daily News*, March 28, 1980.

7 *Malone was too busy*: Thompson, 39.

8 *Johnston was well aware*: Associated Press, "Geologists' Kin Delay Sad Visit," *Eugene Register-Guard*, May 18, 1981.

8 *"We just don't know"*: Ota, Snell, and Zaitz, 8.

10 *It was one of the largest wood-products businesses*: The Weyerhaeuser Company, Tacoma, Washington, 1980 Annual Report.
10 *It owned*: Ibid.
10 *On the basis of its extensive timber holdings*: David Ammons, "Weyerhaeuser Grows Trees and Just Grows," *Seattle Times*, October 20, 1975.
10 *Since becoming president*: Robert Mottram, "Weyerhaeuser . . . Impresses Friends," *Tacoma Tribune*, July 14, 1978.
11 *He had diversified*: The Weyerhaeuser Company, Tacoma, Washington, 1980 Annual Report.
11 *He drove himself to work*: Chet Skreen, "George Weyerhaeuser: Extrovert with a Very Private Core," *Seattle Times*, March 2, 1980.
11 *For vacation*: Thomas J. Murray, "The Mover at Weyerhaeuser," *Dun's*, April 1973, 60.
11 *Weyerhaeuser had been a boxer*: Robert Cantwell, "The Shy Tycoon Who Owns 1/640th of the U.S." *Sports Illustrated*, August 18, 1969, 54.
11 *"There is a quiet power"*: Laura Parker, "Timber Baron Wields a Quiet, Patient Power," *Seattle Post-Intelligencer*, May 29, 1983.
11 *The housing market was collapsing*: Stephen H. Dunphy, "Housing Industry: From Mere Slowdown to a Depression," *Seattle Times*, April 17, 1980.
12 *He could lay off more employees*: Al Watts, "Weyerhaeuser: Good News . . . and Bad," *Seattle Post-Intelligencer*, April 18, 1980.
12 *It wasn't as if*: George H. Weyerhaeuser, "Accent on Regeneration," *American Forests*, January 1973, 23.
13 *"This is not for us"*: Hidy, Hill, and Nevins, 214.
13 *"the best of the S.O.B.'s"*: John G. Mitchell, "Best of the S.O.B.'s," *The Audubon Society Magazine*, September 1974, 49.
15 *From Weyerhaeuser's redbrick mansion*: *Rock Island County: "Where the River Runs West,"* vol. 1 (Moline: The Dispatch and the Rock Island Argus, 2003).
15 *At forty-five, Weyerhaeuser*: Hidy, Hill, and Nevins, 6.
16 *"commenced to dance"*: Healey, 97.
16 *In the spring of 1880, Rock Island*: Elsner, 6.
16 *On the way to his mill*: Quayle and Simpson, passim.
16 *Their lumberyard, along with its associated mills*: Urie, cover illustration.
17 *engaged in a fierce competition*: Healey, 80.
17 *"When I first saw"*: Hill, 14.
17 *they muttered about "aliens"*: Hidy, Hill, and Nevins, 44.
18 *But the Chippewa Falls lumbermen refused*: Fries, 141.
18 *"By 1879 the Chippewa Basin"*: Hidy, Hill, and Nevins, 69.
19 *"I have so many friends"*: Hill, 15.
19 *He was born on November 21*: Healey, 3.
19 *"Probably he wore himself out"*: Hill, 18.
20 *"The voyage"*: Ibid, 22.
20 *"a brewer was often"*: Healey, 30.
21 *"The secret of this"*: Hill, 27.
21 *"quite appalling"*: Healey, 42.
21 *"The happiest days of my life"*: Ibid., 46.
21 *Shortly after the Weyerhaeusers moved*: Hidy, Hill, and Nevins, 5.

22 *He paid his employees on time*: Ibid., 29.
23 *By the 1880s, the nation's railroads*: Hidy, Hill, and Nevins, 135.
23 *But in the latter half*: Cox et al., 128.
23 *In 1867 Weyerhaeuser made his first purchase*: Healey, 62.
23 *"I love the woods life"*: Hidy, Hill, and Nevins, 32.
24 *But an episode from his youth*: Healey, 25.
25 *Americans are, "above all else"*: Cox et al., xv.
25 *In what was to become New England*: Hidy, Hill, and Nevins, 12.
25 *It was the greatest temperate-zone forest*: Cox et al., 62.
25 *So beneficent were the forests*: Cox et al., 5.
25 *"It is a very fine Country"*: Cox, 1.
26 *"a Wooden Town"*: Ibid., xv.
26 *Shipbuilding was New England's most profitable*: Rutkow, 24.
26 *By the end of the 1600s*: Cox et al., 14.
27 *By the time Maine became a state*: Cox, 10.
27 *"seems to be"*: Thoreau, 5.
27 *No New England tree was more prized*: Rutkow, 107.
27 *They tried to outdo*: Edmonds.
27 *"These men cannot live"*: Dwight, vol. 2, 321.
28 *With the possible exception of whaling*: Holbrook, 17.
28 *hung his boots on a nearby tree*: Cox et al., 84.
28 *One logjam on the Kennebec River*: Ibid.
29 *When the colonists landed*: Clawson, 1169.
29 *Today about 40 percent is*: Ibid.
29 *Based on numbers from the US Forest Service*: Ibid., 1171.
29 *But let's say that half*: Ibid.
29 *at least five times the value*: James R. Craig and J. Donald Rimstidt. 1998. "Gold Production History of the United States," *Ore Geology Reviews* 13(6): 407.
30 *The rising water carried away*: Hidy, Hill, and Nevins, 72.
31 *"No other man in America"*: Rutkow, 114.
32 *In St. Paul he bought the mansion*: Healey, 136.
33 *"The timber between the Cascade Range and Puget Sound"*: Martin, 396.
33 *An ingenious backwoods doctor*: Martin, 16.
34 *Strong, gregarious, independent*: Malone, 14.
34 *"two streaks of rust"*: Ibid., 34.
35 *Between 1879 and 1883*: Ibid., 85.
35 *By 1885, four separate transcontinental lines*: White, xxxviii.
35 *At the age of fifty*: Malone, 102.
36 *"Give me enough Swedes"*: Schwantes, *Railroad Signatures,* 133.
36 *So many people walked away*: Hoffman, 109.
36 *L. Frank Baum wrote a story*: Taylor, 413.
37 *In the 1860s and '70s*: Mercer, 6.
37 *The company eventually claimed*: Jensen and Draffan, 7.
38 *The legislation meant*: Ibid., 9.
38 *a hamlet of one hundred people*: Morgan, 43.
39 *Money for wages ran out*: McClelland, *Cowlitz Corridor,* 29.

39 *At one point, unpaid workers*: Crowell, 54.
39 *pitched into the mud*: Kirk and Alexander, 329.
39 *"the only times [I] ever lost money"*: Bureau of Corporations, 186.
40 *At the time, the purchase*: Hidy, Hill, and Nevins, 213.
40 *"There is a great lot of it"*: Ibid., 214.
42 *On May 24, 1935*: Cantwell, 193.
42 *He was wearing*: "Chum of Victim Tells Story: Kidnaped Boy Smiling and Happy on Way Homes, Says Friend," *Seattle Post-Intelligencer*, May 26, 1935.
42 *"How do I get"*: "Weyerhaeuser Kidnapping," HistoryLink: The Free Online Encyclopedia of Washington State History, Essay #7711. Available at http://www.historylink.org.
43 *They crossed a stream*: Cantwell, 194.
44 *"We don't want to hurt anyone"*: "Ransom Note Bared! N.Y. Gang Sought in Kidnapping," *Seattle Post-Intelligencer*, May 27, 1935.
44 *Thousands of people gathered*: "10,000 Pass by Weyerhaeuser Family Home," *Seattle Post-Intelligencer*, May 31, 1935.
44 *They published pictures of George*: "As He Looks Now: Kidnapped Boy Minus Curls," *Seattle Post-Intelligencer*, May 30, 1935.
45 *The next morning, Weyerhaeuser received a telephone call*: Twining, 143.
47 *"I went through all sorts of sensations"*: Cantwell, 211.

PART 2: THE WARNINGS

51 *On Wednesday, March 26*: Beals, Cline, and Koler, 13.
52 *For two decades*: Virginia Culver, "Geologist Predicted St. Helens Eruption," *Denver Post*. April 14, 2009.
52 *they confirmed that the volcano*: Crandell and Mullineaux, C3.
52 *They discovered that Spirit Lake*: Mullineaux and Crandell, 10–11.
52 Potential Hazards from Future Eruptions: Crandell and Mullineaux, C1-C2.
53 *The last eruption was in 1857*: Holmes, 43.
53 *At the meeting in Vancouver, Mullineaux recounted*: Thompson, 33.
53 *Only two people in the city*: Oppenheimer, 34.
54 *Over the years, Weyerhaeuser*: *Karr et al. vs. the State of Washington and Weyerhaeuser Company*, microfilm roll 1, page 1512.
55 *"You mean to tell us"*: Thompson, 34.
55 *"I suppose that is not an unusual phenomenon"*: Brown, 207.
55 *Five years earlier, Mount Baker*: Hodge, Sharp, and Marts, 221–48.
56 *Many of the problems*: Saarinen and Sell, 29–34.
56 *"I'd give them facts"*: Thompson, 42.
57 *But starting on March 31*: Robert Decker, 817.
57 *On April 1, a new eruption*: Foxworthy and Hill, 22.
58 *It had three zones*: Mullineaux, 189.
60 *In the late winter and early spring*: National Oceanic and Atmospheric Administration, *Comparative Climatic Data for the United States Through 2009* (Asheville, NC: National Climatic Data Center), 81.
60 *Some began to sell*: Ota, Snell, and Zaitz, 8.
60 *"If we had to move"*: "'Sightseers go home,' Say Yale, Cougar," *The Lewis County News*, April 9, 1980.

61 *"People are swarming in"*: Foxworthy and Hill, 23.
61 *On March 27, when the first ash clouds*: Sorenson, 162.
61 *Soon maps surfaced*: Ota, Snell, and Zaitz, 8.
65 *the cemeteries were filled with men*: LeMonds, 162.
66 *Most of the money generated*: Robbins, 414.
67 *In 1929, John Killian's father*: "Ralph J. Killian," *Longview Daily News*, March 24, 2010.
69 *"A major deformation like this"*: Rick Seifert, *Longview Daily News*, April 24, 1980.
70 *He and a few of the other geologists*: Associated Press, "St. Helens Like Soviet Volcano," *The Daily World*, April 8, 1980.
70 *Suddenly, on March 30, 1956*: Belousova, Voight, and Belousova, 706.
70 *Early in April, a state snowplow*: Waitt, 50.
71 *Such an eruption could*: Voight, 1695.
72 *"Mount St. Helens has done so many things"*: Foxworthy and Hill, 26.
73 *"I had been there many times"*: Thompson, 75.
73 *An April 25 press release*: Beals, Cline, and Koler, 24.
73 *"The probability of an avalanche"*: United States Department of the Interior, Geological Survey, "Bulge Newest Hazard at Mount St. Helens Volcano" (press release), April 30, 1980.
73 *"would be a pretty extreme scenario"*: Associated Press, "Bulge Poses Massive Slide Danger," *Daily News Journal*, April 29, 1980.
74 *"I have a gut feeling"*: J. Erickson, "Geologist: Bulge May Pop Like Lava Balloon," *News Tribune*, May 6, 1980.
74 *Many had worked previously*: Thompson, 55–71.
75 *"There is a good chance"*: UPI, "Volcano Shows Signs of Eruption," *Hutchinson News*, May 1, 1980.
76 *"There's a hell of a lot"*: Nelson, 214.
76 *Certainly if he ever ran*: Don Jenkins, "Former Cowlitz County Sheriff Les Nelson Dies at 88," February 27, 2010.
77 *"Trying to pin down a geologist"*: Thompson, 36.
77 *"I stuck it out"*: Ibid., 50
78 *He'd been running the Mount St. Helens Lodge*: Rosen, 41.
78 *Truman wouldn't have time*: Donna duBeth, "Sheriff Prepares for Eruptions," *Longview Daily News*, April 29, 1980.
78 *Truman thought the trees*: Rosen, 131.
78 *"The mountain will never hurt me"*: Carson, 34.
79 *"I never wanted to be wrong"*: duBeth, "Sheriff Prepares."
80 *In the spring of 1980*: *Karr et al. v. the State of Washington and Weyerhaeuser Company*, microfilm roll 2, page 843.
80 *"The next log has fallen"*: Kesey, 182–85.
81 *One evening, after setting chokers all day*: Andre Stepankowsky, "No One Stood Taller Than Ralph Killian," *Longview Daily News*, May 29, 2010.
82 *John was making good money*: *Karr et al. vs. the State of Washington and Weyerhaeuser Company*, microfilm roll 2, page 616.
84 *They began by drawing a line*: Ibid., microfilm roll 3, page 257.

85 *So on the western and northwestern sides*: Ibid., 253.
86 *"wanted the zone to extend as far out"*: Ota, Snell, and Zaitz, 13.
86 *"If this isn't Weyerhaeuser and county politics"*: Waitt, 74.
87 *Timber sales had brought in*: Ota, Snell, and Zaitz, 11.
87 *"We were anxious to continue"*: Ibid.
87 *"I missed a lot of sleep"*: Paula Becker, "Ray, Dixy Lee," HistoryLink, November 20, 2004. Available at http://www.historylink.org/index.cfm?DisplayPage=output.cfm&file_id=601.
88 *"There is no evidence"*: Don Duncan, Mark Matassa, and Jim Simon. "Dixy Lee Ray: Unpolitical, Unique, Uncompromising," *The Seattle Times*, January 3, 1994.
88 *She made headlines in Washington, DC*: Rebecca Boren and Debera Carlton Harrell, "Former Gov. Dixy Lee Ray Dies—A Pioneer Amid Controversy," *Seattle Post-Intelligencer*, January 3, 1994.
88 *"It can't be because"*: Lauren Byrnes, "From Mt. Rainier to the Governorship of Washington, Dixy Lee Ray Was a Climber," *AAUW*, October 31, 2013.
88 *During her time as governor*: Ibid.
88 *In 1978 she named a litter*: Ibid.
88 *"I might just read"*: Ota, Snell, and Zaitz, 7.
89 *Toward the end of April*: Lee Seigel, "Gov. Ray May Shut Peak Area," *Longview Daily News*, April 29, 1980.
89 *"You cannot restrict or remove people"*: Peter Lewis et al., "Safety Questions Head for Courts," *Seattle Times*, May 14, 1981.
89 *"It's just impossible"*: Ibid.
89 *Toward the third week of April*: Foxworthy and Hill, 31.
90 *on April 24, the US military*: Mark Bowden, "The Desert One Debacle," *The Atlantic*, May 2006.
90 *The property owners knew*: Thompson, 93.
91 *"It wasn't my jurisdiction"*: Ota, Snell, and Zaitz, 11.
92 *"It was in an area"*: Ibid., 13.
92 *a colleague of Rocky Crandell's had done an analysis*: Thompson, 94.
93 *"in case of violent explosion"*: *Karr et al. vs. the State of Washington and Weyerhaeuser Company*, microfilm roll 2, page 688.
94 *"I wish I did"*: Associated Press, "Harry Truman," *Longview Daily News*, May 15, 1980.

PART 3: THE CONSERVATIONISTS

99 *On August 22, 1986*: diary entries, summer of 1896. Library of Congress, Gifford Pinchot papers, Box 2.
99 *"the most noble looking object"*: Meriwether Lewis diary entries, Sunday, March 30, 1806, available at http://www.gutenberg.org.
100 *His grandfather, Cyrille Pinchot, had emigrated*: Miller, 21.
101 *In that book, published in 1864*: Marsh, passim.
101 *Gifford Pinchot's father*: Miller, 56.
101 *"an amazing question"*: Pinchot, 1.

102 *"first public statement"*: Ibid., 6.
102 *For several months he traveled*: Miller, 78.
102 *"a permanent population"*: Pinchot, 13.
102 *"willing to try"*: Miller, 90.
103 *"If we go on"*: Williams, *Americans and Their Forests*, 397.
103 *Often the timber companies*: Ibid., 236.
103 *"There was a high rate of failure"*: Williams, *Deforesting the Earth*, 306.
104 *"the most important legislation"*: Pinchot, 85.
104 *"the President of the United States may"*: US Congress. "A Bill to Repeal the Timber Culture Laws and for other Purposes." Mar. 3, 1891, ch. 561, 26 Stat. 1095.
104 *Yeoman farmers*: Limerick, 58.
105 *"A national calamity"*: Hidy, Hill, and Nevins, 130.
105 *"the best idea"*: Stegner, 137.
105 *Though many westerners*: Rakestraw, 125.
106 *Of course, the commissioners*: Williams and Miller, 37.
107 *President Cleveland should immediately*: Miller, 136.
107 *Pinchot's study of forestry*: Hays, 42.
108 *"A Republican president will succeed me"*: Hidy, Hill, and Nevins, 139.
108 *"in Washington chiefs"*: Pinchot, 137.
108 *"This is a good place"*: Miller, 143.
108 *"the greatest good"*: Pinchot, 261.
109 *"I had the honor"*: Egan, 36.
109 *the two men chopped wood*: Miller, 150.
109 *"the subject is of importance"*: American Forestry Association, 290.
110 *"at present lumbermen"*: Ibid, 137.
110 *"business view"*: Ibid., 3.
110 *"good business"*: Hays, 42.
110 *"skin the country"*: American Forestry Association, 4.
110 *By the time he left office*: Williams, *Americans and Their Forests*, 421.
111 *"one of the top forests"*: McClure and Mack, 93.
111 *As had been predicted*: Williams, *The U.S. Forest Service in the Pacific Northwest*, 166.
112 *The cut on national forests*: Ibid., 371, Appendix E.
113 *"Multiple use" was always a misnomer*: Hirt, xix.
113 *By the 1970s, ten to twenty*: Ibid., 135.
115 *For decades, groups in Washington and Oregon*: Pryde, 7.
116 *Look at the town*: McClelland, *R. A. Long's Planned City*, passim.
117 *In Saul's clippings file*: "Reader Forum: The Answer on Logging Area Near Mt. St. Helens Is a Loud 'NO!'" *Longview Daily News*, June 28, 1978.
119 *About twenty people were gathered*: Guggenheim, 127–28.
123 *He'd read an article*: Keith Ervin, "St. Helens Calm, but Far from Harmless," *News-Journal*, April 29, 1980.
124 *Nelson tried one last time*: Thompson, 96.
124 *"Your independence and straightforwardness"*: Rosen, 145.
125 *"The only reason I'm here"*: Ibid.
125 *He had painted himself*: Waitt, 80.

126 *"We got the best experts"*: Alan K. Ota, John Snell, and Leslie L. Zaitz, "A Terrible Beauty," *Portland Oregonian*, October 27, 1980, 14.
126 *John and Christy were trying to start a family*: Barbara LaBeo, "A World with Fewer Adventurers, Parents, Researchers," *Longview Daily News*, May 15, 2010.
126 *"They've opened Fawn Lake"*: Waitt, 96.
126 *The Weyerhaeuser Company had told their father*: *Karr et al. vs. the State of Washington and Weyerhaeuser Company*, "Declaration of Ralph Killian," microfilm roll 1, page 1924.
127 *Les Nelson and the other sheriffs*: Ibid., microfilm roll 3, page 571.
127 *The new proposal would*: Memo from Ed Chow to Dixy Lee Ray, May 17, 1980.
127 *Nelson sent a memo*: Memo from Leslie Nelson to Edward Chow, May 15, 1980.
128 *But they delayed for a day*: Waitt, 93.
128 *Also by the third week of May*: Ota, Snell, and Zaitz, 15–16.
128 *Rumors circulated*: Rick Seifert, "Mountain Protesters Get Results," *Longview Daily News*, May 17, 1980.
129 *They loaded leather chairs*: Waitt, 103.
129 *he was planning to come to Castle Rock*: Rosen, 149.
129 *"Oh, c'mon"*: Waitt, 107.
130 *At the staff meeting the geologists*: Thompson, 100.
130 *Johnston had never been to Coldwater II*: Waitt, 101.
132 *"Are you serious?"*: Thompson, 105.
133 *"It's a serious study"*: Rick Seifert, "Geologists Missing: Scientists Likely Died in Mountain Home," *Longview Daily News*, May 20, 1980.
133 *"I'm doing geology at my cabin"*: Waitt, 94.
134 *Earlier on Saturday, Gerry Martin*: Ota, Snell, and Zaitz, 19.
134 *Blackburn was popular*: Jacqui Banaszynski, "A Day for Remembering Reid," *Eugene Register-Guard*, May 30, 1980.
135 *"The whole day I'd just been on edge"*: Steve Twedt, "May 18: Tragic Twists Led to Deaths at Campsites," *Longview Daily News*, August 19, 1981.
135 *Blackburn had talked by phone*: Steve Twedt, "Reid's Story: The Tragedy of One Man's Death," *Longview Daily News*, August 18, 1981.
136 *Northeast of Fawn Lake*: Waitt, 100.
137 *Farthest upriver, Clyde Croft*: Ota, Snell, and Zaitz, 16, 25.
138 *Farthest downriver, six friends*: Waitt, 94.
138 *Since then they had been inseparable: Daily News and Journal-American*, 74.
138 *Earlier that week, the Moores*: Mike and Lu Moore, "May 18th on the Green River," Mount St. Helens Protective Association newsletter.

PART 4: THE ERUPTION

143 *"Do you like to fly?"*: Videotaped interview with Dorothy Stoffel posted by King5 News on May 18, 2010. (Interview no longer available online.)
143 *"They were on their fourth pass"*: Korosec, Rigby, and Stoffel, 5–6.
144 *"Let's get out of here"*: David Ammons, "Eyewitnesses to the Eruption Just 'Microseconds' from Death," *Bremerton Sun*, May 23, 1980.
144 *He redlined the airspeed indicator*: Waitt, 123.

145 *When the cloud hit the resorts*: Voight, 1981, 82–86.
146 *"What are things like up there?"*: Thompson, 108.
146 *The landslide consisted of three blocks*: Moore and Rice, 134.
147 *Technically, the explosion*: Esposti Ongaro et al., 2.
147 *The blast cloud accelerated as it spread*: Kieffer, 379.
147 *his voice was excited, not fearful*: Allan Brettman, "This Is It," *Longview Daily News*, May 17, 1995.
148 *On the wall of his childhood bedroom*: Thompson, 131.
149 *Sunday morning, Pluard*: Alan K. Ota, John Snell, and Leslie L. Zaitz, "A Terrible Beauty," *Portland Oregonian*, October 27, 1980, 23.
149 *Dorothy Stoffel later said*: Waitt, 121.
151 *Three cats traveled with him*: Ota, Snell, and Zaitz, 19.
151 *"Oh oh, I just felt"*: Memorandum from Edward Chow Jr., to Dixy Lee Ray, dated July 29, 1980, recommending that Martin's family be awarded with a plaque, signed "OK, agreed," by Ray.
152 *Martin's microphone switch*: Ota, Snell, and Zaitz, 21.
153 *The next line in his notebook*: Waitt, 155.
153 *The blast cloud reached him*: Twedt, "May 18: Tragic Twists."
154 *Climbers on Mount Rainier*: Waitt, 141.
154 *Before leaving Vader the previous day*: Ross Anderson, "A Father's Lonely Quest," *Seattle Times*, May 10, 1981.
155 *the air pressure suddenly changed*: Rosenbaum and Waitt, 58.
156 *Croft and Handy had camped*: Ota, Snell, and Zaitz, 25.
157 *He pulled out a case*: Ibid.
158 *For three more miles he walked*: Ibid., 27.
159 *"There must be a fire somewhere"*: Ibid., 24.
160 *Bruce wrapped his arms around Sue*: O'Shei, 13.
160 *"Are you okay?"*: Ota, Snell, and Zaitz, 25.
160 *"Bullshit"*: Carol Perkins, "Campers Survive an Ordeal on a Mountain of Death," *Seattle Post-Intelligencer*, May 21, 1980.
160 *"we're not dead yet"*: Waitt, 173.
161 *"If we get out of here alive"*: Ota, Snell, and Zaitz, 25.
161 *His skin hung from his hands*: Waitt, 175.
162 *"Don't leave me here to die!"*: Ota, Snell, and Zaitz, 27.
162 *"Can't you stay with me?"*: Waitt, 176.
163 *"No one could be alive"*: Ibid., 178.
163 *But Brian was no longer at the shack*: Ota, Snell, and Zaitz, 30.
163 *"Hey survivor"*: Ibid. 182.
164 *like a Hardy Boys story*: "Buzz Smith: A Tortuous Hike Through 'Hell on Earth,'" *People*, May 20, 1985, 116.
164 *When they met Danny*: Tom Vogt, "Waking to a Nightmare," *Vancouver Columbian*, April 1, 2010.
165 *Sunday morning, Lu Moore*: Mike and Lu Moore, "May 18th on the Green River," Mount St. Helens Protective Association newsletter.
167 *Shortly before noon*: Mary Don, "Family Saved from Mountain Still Not Home," *Oregon Journal*, May 21, 1980.
168 *"Leave that damn thing"*: Waitt, 326.

PART 5: THE RESCUES

171 *"The mountain's blown"*: Waitt, 135.
172 *Swanson couldn't believe what he was seeing*: Thompson, 109.
172 *After Reber dropped off*: Waitt, 277–79.
174 *"Look at that"*: Alan K. Ota, John Snell, and Leslie L. Zaitz, "A Terrible Beauty," *Portland Oregonian*, October 27, 1980, 28.
174 *"Time to get the hell out of here"*: Waitt, 160.
174 *"He's only doing 80"*: Ibid., 161.
176 *"We are leaving the area!"*: Ota, Snell, and Zaitz, 21.
176 *"a monstrous mural"*: Waitt, 189.
176 *"There it goes"*: Ibid., 213.
176 *he took twenty photographs*: Voight, "Time Scale," 81.
177 *"Run for your lives"*: Waitt, 236.
177 *One strange aspect*: Dewey, passim
178 *In almost every case*: Eisele et al., passim.
180 *As they neared the edge of the water*: Parchman, 49.
181 *"We'd been in the river five minutes"*: Waitt, 200.
181 *"How bad is it?"*: Ibid., 254.
181 *"Camp Baker had [been] built"*: Ibid., 259.
182 *eventually twenty-seven bridges would be destroyed*: "Mount St. Helens—From the 1980 Eruption to 2000," US Geological Survey Fact Sheet 036-00.
182 *"Lots of popping and snapping"*: Waitt, 316.
182 *By the end of the day*: Ibid., 280.
183 "El volcán esta explotando": Parchman, 23.
183 *"It got hot right away"*: Waitt, 164.
183 *he later would be found*: Ota, Snell, and Zaitz, 32.
183 *When the flight surgeon reached out*: Ibid., 29.
184 *One entrepreneur used a crane*: John O'Ryan, "Logjam Brings Kelso a Bonanza," *Seattle Post-Intelligencer*, May 19, 1980.
184 *Weyerhaeuser began suing*: "A Suit to Get Logs Back," *Seattle Post-Intelligencer*, July 21, 1980.
184 *The flood carried immense amounts*: Schuster, 705.
185 *"You see the smoke"*: Waitt, 125.
186 *town of Othello*: Personal communication, Edna Taylor.
186 *emergency runs to convenience stores*: Personal communication, Rick Olson.
186 *to collect the ash in jars*: Personal communication, Dave Olson.
187 *Even a year later*: Waitt, 345.
188 *"This guy looks awfully young"*: Ota, Snell, and Zaitz, 32.
188 *But below the ash*: Ibid., 31.
189 *"If you don't let these kids"*: Waitt, 338.
192 *"This is all very interesting"*: Dick Johnston, "Carter Vows Fast Volcano Action," *The Oregonian*, May 22, 1980.
193 *"not so much for the people who were killed"*: "President Arrives to Assess the Volcano Damage," *Seattle Post-Intelligencer*, May 22, 1980.
193 *"Don't ask me about the governor"*: Lee Siegel, "Carter Calls Volcano Disaster," *Bellevue Journal American*, May 22, 1980.

193 *"My headphones don't work"*: Waitt, 335.
194 *"Many people chose to remain"*: Lee Siegel, "Dixy: Victims Chose to Stay; We're Not to Blame," *Bellevue Journal American*, May 21, 1980.
194 *"One of the reasons for the loss of life"*: Ota, Snell, and Zaitz, 34.
194 *"Fearing the worst, state officials"*: Findley, 20
195 *"It's like, you can't be half pregnant": Karr et al. vs. the State of Washington and Weyerhaeuser Company*, microfilm roll 2, page 1166.

PART 6: THE MONUMENT

202 *Then the association got a break*: Mount St. Helens Protective Association, unpublished manuscript.
202 *The previous month, Weyerhaeuser*: Pam Leven, "Weyerhaeuser Begins Salvage of Volcano-Damaged Timber," *Seattle Post-Intelligencer*, September 16, 1980.
203 *"Land management decisions"*: Parchman, 251.
203 *"It wasn't just the volcano"*: Nance, 6.
204 *a group of angry westerners*: Short, 8.
204 *Actually, very similar movements*: Ibid., 11.
204 *"I renew my pledge"*: Ibid., 9.
206 *Mount St. Helens was withdrawn*: Rick Seifer, "Carter Eyes Peak as Monument," *Longview Daily News*, December 30, 1980.
209 *In the years before her death*: Knute Berger, "A Water Taxi Named 'Dixy'?" *Crosscut*, August 5, 2014.
211 *In hearings on the proposed bills*: United States Congress, Mount St. Helens National Volcanic Area: Joint Hearings Before the Subcommittee on Forests, Family Farms, and Energy of the Committee on Agriculture and the Subcommittee on Public Lands and National Parks of the Committee on Interior and Insular Affairs, House of Representatives, Ninety-seventh Congress, Second Session, on H.R. 5281, H.R. 5773, H.R. 5787, March 11, 1982, Washington, DC, April 3, 1982, Vancouver, WA.
214 *It and Burlington Northern traded away*: "Peak Bill's Fate Up to President," *Longview Daily News*, August 18, 1982.
216 *"What is the real motive"*: Short, 52.
216 *"too small to be a state"*: Patricia Sullivan, "Anne Gorsuch Burford, 62, Dies: Reagan EPA Director," *Washington Post*, July 22, 2004.
216 *The Office of Management and Budget*: Ward Sinclair, "White House Still Wary of Peak Monument Land Swap," *Longview Daily News*, August 2, 1982.
217 *On August 26, 1982*: Andre Stepankowsky, "Reagan Signs Monument Bill," *Longview Daily News*, August 27, 1982.
217 *By the end of his presidency*: Steven F. Hayward, "Ronald Reagan and the Environment," *inFocus Quarterly*, Fall 2009.

PART 7: DECLINE AND RENEWAL

221 *"from April 1980, until May 18": Karr et al. vs. the State of Washington and Weyerhaeuser Company*, microfilm roll 1, page 1517.
222 *"Weyerhaeuser did not vary the work location"*: Parchman, 297.

222 *"They chose to ignore that"*: Ota, Snell, and Zaitz, 13.
222 *The majority of the jurors*: Larry Lange, ""Act of God' Factor Splits the Jury in Volcano Suit," *Seattle Post-Intelligencer*, December 3, 1985.
223 *Finally the families agreed to settle*: Larry Lange, "St. Helens Suit Against Weyerhaeuser Settled," *Seattle Post-Intelligencer*, February 7, 1987.
223 *"I'm glad it's over"*: Parchman, 308.
223 *Dixy Lee Ray and George Weyerhaeuser certainly talked*: Timothy Egan, "Dixy Changes Story on Red Zone Limits," *Seattle Post-Intelligencer*, March 10, 1984.
223 *"Before May 18 . . . I spoke several times"*: *Karr et al. vs. the State of Washington and Weyerhaeuser Company*, microfilm roll 2, page 296.
223 *As Weyerhaeuser said in his affidavit*: Ibid., 279–80.
224 *The age of heroic, highball*: Penttila and Bertroch, 96.
224 *Industrial timber harvests*: Data downloaded from Washington State Department of Natural Resources. Available at http://www.dnr.wa.gov/TimberHarvestReports.
224 *Soon after the salvage operation ended*: LeMonds, 122.
225 *"We're not a philanthropic enterprise"*: George Draffan, "A Profile of the Weyerhaeuser Corporation." Available at http://www.endgame.org/weyerprofile.html.
226 *dropped by one-third*: Warren, 3.
226 *An obscure Massachusetts environmental group*: Dietrich, 91.
226 *"a school for the young"*: LeMonds, 101.
227 *In 1980, Weyerhaeuser had 48,000 employees*: Employment numbers are from annual reports and from http://www.weyerhaeuser.com/Sustainability/DataAndGRIIindex.
227 *he was the kind of guy*: Recounted at the memorial service for George Weyerhaeuser, Jr., held on May 31, 2013, at the Museum of Glass in Tacoma.
228 *none of the company's top corporate officers*: Bill Virgin, "Local Family Ties That Bind to Companies Evolve Through the Years," *News Tribune*, April 28, 2013.
228 *"You feel like you want to"*: Associated Press, "Families Continue Search for Mt. St. Helen's Victims," *Winnipeg Free Press*, May 14, 1981.
228 *"I don't know if it was an illusion"*: Parchman, 335.
229 *"We'd failed"*: Waitt, 323.
229 *"Naturally, if I had it all to do over"*: Associated Press, "Scientists Frankly Admit Volcano Confounded Them," *Bremerton Sun*, May 22, 1980.
230 *"Mount St. Helens has probably included"*: Crandell, Mullineaux, and Rubin, 439.
230 *"strong laterally directed explosion"*: Crandell and Mullineaux, C4.
230 *they have improved their procedures*: Saarinen and Sell, 52.
231 *"would have preferred the simple hazards"*: Rowley et al., 7.
231 *It also created a standardized*: Fearnley et al., passim.
231 *a fantastic suite of new technologies*: Vallance et al., passim.
232 *geologists around the world have identified hundreds*: Tilling, "Mount St. Helens 20 Years Later," 16.
232 *Mount Augustine in Alaska*: Siebert, 53.
232 *Mount Rainier, had shed prodigious amounts*: Hoblitt et al., 6.
233 *Today, hundreds of thousands of people*: Driedger and Scott, 12.

233 *with a significant eruption occurring*: Kiver, 8.
233 *When Mount Mazama in southern Oregon erupted*: Harris, 133–56.
233 *If the caldera under Yellowstone National Park*: Lowenstern et al.
233 *In 1985 a relatively small eruption*: Tilling, "Volcanic Hazards and Their Mitigation," 252–54.
233 *The enormous 1991 eruption of Mount Pinatubo*: Fisher, Heiken, and Hulen, 53.
234 *an average of fifty to sixty volcanoes*: Peterson, 4161.
234 *about twenty separate eruptions*: Tilling, Topinka, and Swanson, 30.
234 *From 2004 until 2008 the volcano erupted again*: Sherrod, Scott, and Stauffer, 6.
236 *Studies conducted in the monument*: Dale, Swanson, and Crisafulli, passim.
236 *One surprising observation*: Franklin and MacMahon, 1183.
236 *more than 120 kinds of spiders*: Lynne Peeples, "11 Surprising Natural Lessons from Mount St. Helens," *Scientific American*, May 19, 2010. Available at http://www.scientificamerican.com/article/mount-st-helens-lessons.
236 *An especially important plant*: Colasurdo, 247.
237 *The succession of species*: Nash, 572.
237 *an expensive and controversial aerial reseeding program*: Carson, 115.
238 *the area surrounding Mount St. Helens today*: Le Guin, 5.
239 *In 1985 she and a group of other environmentalists*: From the mission statement of the Gifford Pinchot Task Force: http://www.gptaskforce.org/about.
240 *The lake water splashed*: Decker, *The Rebirth of Mount St. Helens*, 26.
240 *The Forest Service is a different agency*: Dietrich, 312–13.
240 *Between 1982 and 2008*: "Mount St. Helens National Volcanic Monument: A Summary of Economic Performance in the Surrounding Communities," Headwaters Economics, 2011.
240 *Hunters, mountain bikers*: Cooper, 5.
240 *sold its holdings in the area*: Tom Paulu, "Coalition Strives to Reopen High Lakes Area to Public," *Longview Daily News*, September 27, 2012.
240 *An even greater threat*: Natalie St. John, "Ascot CEO: Mining Gamble Near Volcano Could Bring Thousands of Jobs," *Longview Daily News*, December 8, 2012. A subsequent series of articles by St. John explored the mining proposals and consequences in detail.

EPILOGUE

243 *Since 1980, geologists have learned*: Doughton, passim.
243 *Computer simulations of a large earthquake*: Ibid., 85.
243 *According to a 2006 study*: Ripley, xiv.

BIBLIOGRAPHY

American Forestry Association. 1905. *Proceedings of the American Forest Congress Held at Washington, D.C., January 2 to 6, 1905.* Washington, DC: H. M. Suter Publishing Co.

Anderson, David A. 2013. *Images of America: Mount St. Helens.* Charleston, SC: Arcadia Publishing.

Beals, Herbert K., Scott Cline, and Julie M. Koler. 1981. *On the Mountain's Brink: A Forest Service History of the 1980 Mount St. Helens Volcanic Emergency.* Washington, DC: US Department of Agriculture.

Belousova, Alexander, Barry Voight, and Marina Belousova. 2007. "Directed Blasts and Blast-Generated Pyroclastic Density Currents: A Comparison of the Bezymianny 1956, Mount St Helens 1980, and Soufrière Hills, Montserrat 1997 Eruptions and Deposits." *Bulletin of Volcanology* 69 (7): 701–40.

Berger, Knute. 2009. *Pugetopolis: A Mossback Takes on Growth Addicts, Weather Wimps, and the Myth of Seattle Nice.* Seattle: Sasquatch Books.

Brown, Jerry J. 1982. "Role of the U.S. Forest Service at Mount St. Helens." In *Status of Volcanic Prediction and Emergency Response Capabilities in Volcanic Hazard Zones of California*, edited by Roger C. Martin and James F. Davis, 201–12. Sacramento: California Department of Conservation.

Bureau of Corporations. 1913. *The Lumber Industry, Part 1: Standing Timber.* Washington, DC: Government Printing Office.

Cantwell, Robert. 1972. *The Hidden Northwest,* New York: J. B. Lippincott.

Carson, Rob. 2000. *Mount St. Helens: The Eruption and Recovery of a Volcano.* Revised edition. Seattle: Sasquatch Books.

Christiansen, Robert L., and Donald W. Peterson. 1981. "Chronology of the 1980 Eruptive Activity." In *The 1980 Eruptions of Mount St. Helens, Washington,* edited by Peter W. Lipman and Donal R. Mullineaux, 17–30. Geological Survey Professional Paper 1250. Washington, DC: Government Printing Office.

Clawson, Marion. 1979. "Forests in the Long Sweep of American History." *Science* 204: 1168–174.

Colasurdo, Christine. 2010. *Return to Spirit Lake: Life and Landscape at Mount St. Helens*. Astoria, OR: Radiolarian Press.

The Columbian. 1980. *Mount St. Helens Holocaust: A Diary of Destruction*. Lubbock, TX: C. F. Boone Publishers.

Cooper, Ryan, with Eli Lehrer. 2013. "The Economic Benefits of Protected Lands: A Government-Life Approach." R. Street Policy Study No. 12. Washington, DC: R Street Institute.

Cox, Thomas R. 1974. *The Lumberman's Frontier: Three Centuries of Land Use, Society, and Change in America's Forests*. Corvallis: Oregon State University Press.

Cox, Thomas R., Robert S. Maxwell, Phillip Drennon Thomas, and Joseph J. Malone. 1985. *This Well-Wooded Land: Americans and Their Forests from Colonial Times to the Present*. Lincoln: University of Nebraska Press.

Crandell, Dwight R., and Donal R. Mullineaux. 1978. *Potential Hazards from Future Eruptions of Mount St. Helens Volcano*. Geological Survey Bulletin 1383-C. Washington, DC: Government Printing Office.

Crandell, D. R., D. R. Mullineaux, and M. Rubin. 1975. "Mount St. Helens Volcano—Recent and Future Behavior." *Science* 187: 438–41.

Cronon, William, ed. 1995. *Uncommon Ground: Toward Reinventing Nature*. New York: Norton.

Crowell, Sandra A. 2007. *The Land Called Lewis: A History of Lewis County, Washington*. Centralia, WA: Gorham Printing.

Daily News and Journal-American. 1980. *Volcano: The Eruption of Mount St. Helens*. Longview, WA: Longview Publishing Company.

Dale, Virginia H., Frederick J. Swanson, and Charles M. Crisafulli, eds. 2005. *Ecological Responses to the 1980 Eruption of Mount St. Helens*. New York: Springer-Verlag.

Decker, Barbara. 2007. *The Rebirth of Mount St. Helens*. Mariposa, CA: Sierra Press.

Decker, Robert W. 1981. "The 1980 Activity—A Case Study in Forecasting Volcanic Eruptions." In *The 1980 Eruptions of Mount St. Helens, Washington*, edited by Peter W. Lipman and Donal R. Mullineaux, 815–20. Geological Survey Professional Paper 1250. Washington, DC: Government Printing Office.

Decker, Robert, and Barbara Decker. 1981. "The Eruptions of Mount St. Helens." *Scientific American* (May).

Dewey, John M. 1985. "The Propagation of Sound from the Eruption of Mt. St. Helens on 18 May 1980." *Northwest Science* 59 (2): 79–92.

Dietrich, William. 2010. *The Final Forest: The Battle for the Last Great Trees of the Pacific Northwest*. Paperback edition, with new preface and afterword. Seattle: University of Washington Press.

Doughton, Sandi. 2013. *Full Rip 9.0: The Next Big Earthquake in the Pacific Northwest*. Seattle: Sasquatch Books.

Driedger, Carolyn, and William E. Scott. 2008. *Mount Rainier—Living Safely with a Volcano in Your Backyard*. Fact Sheet 2008–3062. Reston, VA: US Geological Survey.

Durbin, Kathie. 1996. *Tree Huggers: Victory, Defeat, and Renewal in the Northwest Ancient Forest Campaign*. Seattle: The Mountaineers.

Dwight, Timothy. 1821–22. *Travels; in New-England and New-York*. New Haven: Timothy Dwight.

Dzurisin, Dan, Peter H. Stauffer, and James W. Hendley. 2008. *Living with Volcanic Risk in the Cascades*. USGS Fact Sheet 165–97. Reston, VA: US Geological Survey.

Egan, Timothy. 2009. *The Big Burn: Teddy Roosevelt and the Fire that Saved America*. Boston: Houghton Mifflin Harcourt.

Edmonds, Michael. 2009. *Out of the Northwoods: The Many Lives of Paul Bunyan, with More than 100 Logging Camp Tales*. Madison: Wisconsin Historical Society.

Ehrlich, Paul R., Carl Sagan, Donald Kennedy, and Walter Orr Roberts. 1984. *The Cold and the Dark: The World After Nuclear War*. New York: Norton.

Eisele, John W., Ronald L. O'Halloran, Donald T. Reay, George R. Lindholm, Larry V. Lewman, and William J. Brady. 1981. "Deaths During the May 18, 1980, Eruption of Mount St. Helens." *New England Journal of Medicine* 305: 931–36.

Elsner, B J., ed. 1988. *Rock Island: Yesterday, Today & Tomorrow*. Montezuma, IA: Sutherland Printing Co.

Endo, Elliott T., Stephen D. Malone, Linda L. Noson, and Craig S. Weaver. 1981. "Locations, Magnitudes, and Statistics of the March 20–May 18 Earthquake Sequence." In *The 1980 Eruptions of Mount St. Helens, Washington*, edited by Peter W. Lipman and Donal R. Mullineaux, 93–107. Geological Survey Professional Paper 1250. Washington, DC: Government Printing Office.

Esposti Ongaro, T., A. B. Clarke, B. Voight, A. Neri, and C. Widiwijayanti. 2012. "Multiphase flow dynamics of pyroclastic density currents during the lateral blast of Mount St. Helens." *Journal of Geophysical Research* 117 (B6): 1–22.

Fairfield, Clara. 1981. "The Acoustic Effects of the Eruption of Mount St. Helens May 18, 1980." In *Mount St. Helens—One Year Later*, edited by A. C. Keller, 83–85. Cheney, WA: Eastern Washington University Press.

Fallows, James. 2000. "Saving Salmon or Seattle?" *Atlantic Monthly* (October), 20–26.

Fang, Janet. 2010. "Hot Science from a Volcanic Crisis." *Nature* 465: 146–47.

Fearnley, C. J., W. J. McGuire, G. Davies, and J. Twigg. 2012. "Standardisation of the USGS Volcano Alert Level System: analysis and ramifications." *Bulletin of Volcanology* 72: 2023–36.

Ficken, Robert E. 1979. "Weyerhaeuser and the Pacific Northwest Timber Industry, 1899–1903." *Pacific Northwest Quarterly* 70: 146–54.

Ficken, Robert E. 1987. *The Forested Land: A History of Lumbering in Western Washington*. Seattle: University of Washington Press.

Findley, Rowe. 1981. "St. Helens: Mountain with a Death Wish," *National Geographic* (January), 3–65.

Fisher, Richard V., Grant Keiken, and Jeffrey B. Hulen. 1997. *Volcanoes: Crucibles of Change*. Princeton, NJ: Princeton University Press.

Foxworthy, Bruce L., and Mary Hill. 1982. *Volcanic Eruptions of 1980 at Mount St. Helens: The First 100 Days*. USGS Professional Paper 1249. Washington, DC: US Geological Survey.

Francis, Peter, and Clive Oppenheimer. 2003. *Volcanoes*. New York: Oxford University Press.

Franklin, Jerry F., and James A. MacMahon. 2000. "Messages from a Mountain." *Science* 288:1183–184.

Fries, Robert F. 1951. *Empire in Pine: The Story of Lumbering in Wisconsin*. Madison: The State Historical Society of Wisconsin.

Goodrich, Charles, Kathleen Dean Moore, and Frederick J. Swanson, eds. 2008. *In the Blast Zone: Catastrophe and Renewal on Mount St. Helens*. Corvallis: Oregon State University Press.

Guggenheim, Alan. 1986. *Spirit Lake People: Memories of Mount St. Helens*. Gresham: Salem Press.

Harris, Stephen L. 2005. *Fire Mountains of the West: The Cascade and Mono Lake Volcanoes*. Third edition. Missoula, MT: Mountain Press.

Hays, Samuel. 2009. *The American People and the National Forests: The First Century of the U.S. Forest Service*. Pittsburgh: University of Pittsburgh Press.

Healey, Judith Koll. 2013. *Frederick Weyerhaeuser and the American West*. St. Paul: Minnesota Historical Society Press.

Hidy, Ralph W., Frank Earnest Hill, and Allan Nevins. 1963. *Timber and Men: The Weyerhaeuser Story*. New York: The Macmillan Company.

Hill, William Bancroft. 1940. *Frederick Weyerhaeuser, Pioneer Lumberman*. Minneapolis: The McGill Lithographic Company.

Hirt, Paul W. 1994. *Conspiracy of Optimism: Management of the National Forests since World War Two*. Lincoln: University of Nebraska Press.

Hoblitt, R. P., J. S. Walder, C. L. Driedger, K. M. Scott, P. T. Pringle, and J. W. Vallance. 1998. *Volcano Hazards from Mount Rainier, Washington*. Open-File Report 98–428. Reston, VA: US Geological Survey.

Hodge, David, Virginia Sharp, and Marion Marts. 1979. "Contemporary Responses to Volcanism: Case Studies from the Cascades and Hawaii." In *Volcanic Activity and Human Ecology*, edited by Payson D. Sheets and Donald K. Grayson, 221–48. New York: Academic Press.

Hoffman, Charles. 1970. *The Depression of the Nineties: An Economic History*. Westport, CT: Greenwood Publishing.

Holbrook, Steward. 1956. *Holy Old Mackinaw: A Natural History of the American Lumberjack*. New York: Macmillan.

Holmes, Kenneth L. 1980. *Mount St. Helens: Lady with a Past*. Salem, OR: Salem Press.

Jensen, Derrick, and George Draffan. 1995. *Railroads and Clearcuts: Legacy of Congress's 1864 Northern Pacific Railroad Land Grant*. Spokane: Inland Empire Public Lands Council.

Jones, Alden. 1974. *From Jamestown to Coffin Rock: A History of Weyerhaeuser Operations in Southwest Washington*. Tacoma: Weyerhaeuser Company.

Kelso, Linda. 1980. *Volcano: First Seventy Days: Mount St. Helens, 1980*. Beaverton, OR: Beautiful America Publishing Co.

Kendrick, T. D. 1956. *The Lisbon Earthquake*. London: Methuen and Co.

Kesey, Ken. 1964. *Sometimes a Great Notion*. New York: Viking Penguin.

Kieffer, Susan Werner. 1981. "Fluid Dynamics of the May 18 Blast at Mount St. Helens." In *The 1980 Eruptions of Mount St. Helens, Washington*, edited by Peter W. Lipman and Donal R. Mullineaux, 379–400. Geological Survey Professional Paper 1250. Washington, DC: Government Printing Office.

Kirk, Ruth, and Carmela Alexander. 1995. *Exploring Washington's Past: A Road Guide to History*. Revised edition. Seattle: University of Washington Press.

Kiver, Eugene P. 1982. "The Cascade Volcanoes: Comparison of Geologic and Historic Records." In *Mount St. Helens: One Year Later*, edited by S. A. C. Keller, 3–12. Cheney, WA: Eastern Washington University Press.

Kline, Benjamin. 2011. *First Along the River: A Brief History of the U.S. Environmental Movement*. Fourth edition. Lanham, MD: Rowman and Littlefield.

Klingle, Matthew. 2007. *Emerald City: An Environmental History of Seattle*. New Haven: Yale University Press.

Koenninger, Tom, ed. 1980. *Mount St. Helens Holocaust: A Diary of Destruction*. Lubbock, Texas: C. F. Boone Publishers.

Korosec, M. A., James G. Rigby, and Keith L. Stoffel. 1980. *The 1980 Eruption of Mount St. Helens, Washington. Part I, March 20–May 19, 1980*. Information Circular 71. Olympia: Washington State Department of Natural Resources, Division of Geology and Earth Resources.

LeMonds, James. 2001. *Deadfall: Generations of Logging in the Pacific Northwest*. Missoula, MT: Mountain Press.

Le Guin, Ursula K. 1983. *In the Red Zone*. Northridge, CA: Lord John Press.

———. "Coming Back to the Lady." In *In the Blast Zone: Catastrophe and Renewal on Mount St. Helens*, edited by Charles Goodrich, Kathleen Dean Moore, and Frederick J. Swanson, 2–9. Corvallis: Oregon State University Press.

Lien, Carsten. 1991. *Olympic Battleground: The Power Politics of Timber Preservation*. San Francisco: Sierra Club Books.

Limerick, Patricia Nelson. 1987. *The Legacy of Conquest: The Unbroken Past of the American West*. New York: Norton.

Lipman, Peter W., and Donal R. Mullineaux, eds. 1981. *The 1980 Eruptions of Mount St. Helens, Washington*. USGS Professional Paper 1250. Washington, DC: Government Printing Office.

Lowenstern, Jacob B., Robert L. Christiansen, Robert B. Smith, Lisa A. Morgan, and Henry Heasler. 2005. *Steam Explosions, Earthquakes, and Volcanic Eruptions—What's in Yellowstone's Future?* USGS Fact Sheet 2005–3024. Reston, VA: US Geological Survey.

Lowenthal, David. 2000. *George Perkins Marsh: Prophet of Conservation*. Seattle: University of Washington Press.

Luis, Michael. 2012. *Century 21 City: Seattle's Fifty Year Journey from World's Fair to World Stage*. Medina, WA: Fairweather Publishing.

Macaulay, Tom. 1977. "Mount St. Helens: Beautiful Mountain of Death." *Northwest Magazine* (January 23), 29.

MacCleery, Douglas. 1992. *American Forests: A History of Resiliency and Recovery*. Washington, DC: US Department of Agriculture.

Malone, Michael P. 1996. *James J. Hill: Empire Builder of the Northwest*. Norman: University of Oklahoma Press.

Manes, Stephen, and Paul Andrews 1992. *Gates: How Microsoft's Mogul Reinvented an Industry and Made Himself the Richest Man in America*. New York: Doubleday.

Marsh, George Perkins. 1865. *Man and Nature; or, Physical Geography as Modified by Human Action*. New York: Charles Scribner.

Martin, Albro. 1976. *James J. Hill and the Opening of the Northwest*. New York: Oxford University Press.

Mater, Jean. 1997. *Reinventing the Forest Industry*. Wilsonville, OR: GreenTree Press.

McClelland, John M. Jr. 1976. *R. A. Long's Planned City: The Story of Longview*. Longview, WA: Longview Publishing.

——. 1984. *Cowlitz Corridor: Historical River Highway of the Pacific Northwest*. Longview: Longview Publishing Co.

McClure, Rick, and Cheryl Mack. 1999. *"For the Greatest Good": Early History of Gifford Pinchot National Forest*. Seattle: Northwest Interpretive Association.

Meadows, Donella H., Dennis L. Meadows, Jorgen Randers, and William W. Behrens III. 1972. *The Limits to Growth*. New York: New American Library.

Mercer, Lloyd J. 1982. *Railroads and Land Grant Policy: A Study in Government Intervention*. New York: Academic Press.

Miles, John C. 2009. *Wilderness in National Parks: Playground or Preserve*. Seattle: University of Washington Press.

Miller, Char. 2001. *Gifford Pinchot and the Making of Modern Environmentalism*. Washington, DC: Island Press.

Moody, Fred. 2004. *Seattle and the Demons of Ambition: From Boom to Bust in the Number One City of the Future*. New York: St. Martin's.

Moore, James G., and Carl J. Rice. 1984. "Chronology and Character of the May 18, 1980, Explosive Eruptions of Mount St. Helens." In *Explosive Volcanism: Inception, Evolution, and Hazards*, 133–42. Washington, DC: National Academies Press.

Moore, Kathleen Dean. 2009. "In the Shadow of the Cedars: Spiritual Values of Old-Growth Forests." In *Old Growth in a New World: A Pacific Northwest Icon Reexamined*, edited by Thomas A. Spies and Sally L. Duncan, 168–75. Washington, DC: Island Press.

Morgan, Murray. 1984. *South on the Sound: An Illustrated History of Tacoma and Pierce County*. Woodland Hills, CA: Windsor Publications.

Mullineaux, Donal R. 1982. In *Status of Volcanic Prediction and Emergency Response Capabilities in Volcanic Hazard Zones of California*, edited by Roger C. Martin and James F. Davis, 187–91. Sacramento: California Department of Conservation.

Mullineaux, Donal R., and Dwight R. Crandell. 1981. "The Eruptive History of Mount St. Helens." In *The 1980 Eruptions of Mount St. Helens, Washington*, edited by Peter W. Lipman and Donal R. Mullineaux, 3–15. Geological Survey Professional Paper 1250. Washington, DC: Government Printing Office.

Nance, Ancil. 1982. "Logging in the Red Zone." *American Forests* 88 (April): 6–10, 57-58.

Nash, Steve. 2010. "Making Sense of Mount St. Helens." *BioScience* 60 (September): 571–75.

Nelson, Les. 1982. "Local Government's Experience at Mount St. Helens." In *Status of Volcanic Prediction and Emergency Response Capabilities in Volcanic Hazard Zones of California*, edited by Roger C. Martin and James F. Davis, 213–17. Sacramento: California Department of Conservation.

Noah, Timothy. 2012. *The Great Divergence: America's Growing Inequality Crisis and What We Can Do About It.* New York: Bloomsbury.

Nokes, J. Richard et al. 1980. *Mount St. Helens: The Volcano.* Portland, OR: The Oregonian Publishing Co.

Norse, Elliott A. et al. 1990. *Ancient Forests of the Pacific Northwest.* Washington, DC: Island Press.

Nur, Amos. 2008. *Apocalypse: Earthquakes, Archaeology, and the Wrath of God.* Princeton, NJ: Princeton University Press.

Oppenheimer, Clive. 2011. *Eruptions that Shook the World.* New York: Cambridge University Press.

O'Shei, Tim. 2007. *Volcanic Eruption: Susan Ruff and Bruce Nelson's Story of Survival.* Mankato, MN: Capstone Press.

Ota, Alan K., John Snell, and Leslie L. Zaitz. 1980. "A Terrible Beauty," *Portland Oregonian*, October 27.

Parchman, Frank. 2005. *Echoes of Fury: The 1980 Eruption of Mount St. Helens and the Lives It Changed Forever.* Kenmore, WA: Epicenter Press.

Penttila, Bryan, and Karen Bertroch. 2011. *When Logging Was Logging: 100 Years of Big Timber in Southwest Washington.* Virginia Beach, VA: Donning Company Publishers.

Perry, Ronald W., and Michael K. Lindell. 1990. *Living with Mount St. Helens.* Pullman: Washington State University Press.

Peterson, Donald W. 1988. "Volcanic Hazards and Public Response." *Journal of Geophysical Research* 93 (B5): 4161–170.

Pinchot, Gifford. 1947. *Breaking New Ground.* New York: Harcourt Brace Jovanovich.

Pinsker, Lisa M. 2004. "Paths of Destruction: The Hidden Threat at Mount Rainier." *Geotimes* 49 (April): 16–23.

Proctor, James D. 1996. "Whose Nature? The Contested Moral Terrain of Ancient Forests." In *Uncommon Ground: Toward Reinventing Nature*, edited by William Cronon, 269–97. New York: Norton.

Pringle, Patrick T. 1993. *Roadside Geology of Mount St. Helens National Volcanic Monument and Vicinity.* Olympia, WA: Washington Department of Natural Resources.

Pryde, Philip R. 1968. "Mount St. Helens: A Possible National Monument." *National Parks* (May): 7–10.

Pyle, Robert Michael. 1986. *Wintergreen: Rambles in a Ravaged Land.* New York: Scribner.

Quayle, W., and H. Simpson. 1888. *Rock Island Illustrated. A Brief Review of a Busy City in the Mississippi Valley.* Rock Island, IL: Daily Argus Press.

Raban, Jonathan. 2010. *Driving Home: An American Journey.* New York: Pantheon.

Rajala, Richard A. 1998. *Clearcutting the Pacific Rain Forest: Production, Science and Regulation.* Vancouver: University of British Columbia Press.

Rakestraw, Lawrence. 1953. "Uncle Sam's Forest Reserves." *Pacific Northwest Quarterly* (October).

Raphael, Ray. 1981. *Tree Talk: The People and Politics of Timber.* Washington, DC: Island Press.

Robbins, William G. 1987. "Lumber Production and Community Stability: A View from the Pacific Northwest." *Journal of Forest History* 31 (October): 187–96.

Rosen, Shirley. 1981. *Truman of St. Helens: The Man and His Mountain*. Bothell, WA: Rosebud Publishing.

Rosenbaum, J. G., and Richard B. Waitt Jr. 1981. "Summary of Eyewitness Accounts of the May 18 Eruption." In *The 1980 Eruptions of Mount St. Helens, Washington*, edited by Peter W. Lipman and Donal R. Mullineaux, 53–67. Geological Survey Professional Paper 1250. Washington, DC: Government Printing Office.

Rosenfeld, Charles, and Robert Cooke. 1982. *Earthfire: The Eruption of Mount St. Helens*. Cambridge, MA: MIT Press.

Rowley, Peter D., M. H. Hait Jr., Donald R. Finley, Donovan B. Kelly, Susan L. Russell-Robinson, Jane M. Buchanan-Banks, Katherine V. Cashman, and Edna G. King. 1984. *Between Mount St. Helens and the World: How the U.S. Geological Survey Provided News-Media Information on the 1980 Volcanic Eruptions*. US Geological Survey Circular 921. Alexandria, VA: US Geological Survey.

Runte, Alfred. 1983. "Burlington Northern and the Legacy of Mount Saint Helens." *Pacific Northwest Quarterly* 74 (July): 116–23.

Rutkow, Eric. 2012. *American Canopy: Trees, Forests, and the Making of a Nation*. New York: Scribner.

Saarinen, Thomas F., and James L. Sell. 1985. *Warning and Response to the Mount St. Helens Eruption*. Albany: State University of New York Press.

Sale, Roger. 1976. *Seattle: Past to Present*. Seattle: University of Washington Press.

Scarth, Alwyn. 2009. *Vesuvius: A Biography*. Princeton, NJ: Princeton University Press.

Schellenberger, Michael, and Ted Nordhaus. 2004. *The Death of Environmentalism: Global Warming Politics in a Post-Environmental World*. Available at www.thebreakthrough.org.

Schultz, Howard. 1997. *Pour Your Heart into It: How Starbucks Built a Company One Cup at a Time*. New York: Hyperion.

Schwantes, Carlos A. 1993. *Railroad Signatures Across the Pacific Northwest*. Seattle: University of Washington.

———. 1996. *The Pacific Northwest: An Interpretive History*. Lincoln: University of Nebraska Press.

Schuster, Robert L. 1981. "Effects of the Eruptions on Civil Works and Operations in the Pacific Northwest." In *The 1980 Eruptions of Mount St. Helens, Washington*, edited by Peter W. Lipman and Donal R. Mullineaux, 701–18. Geological Survey Professional Paper 1250. Washington, DC: Government Printing Office.

Sensel, Joni. 1999. *Traditions Through the Trees: Weyerhaeuser's First 100 Years*. Seattle: Documentary Book Publishers.

Sherrod, D. R., W. E. Scott, and P. H. Stauffer, eds. 2008. *A Volcano Rekindled: The Renewed Eruption of Mount St. Helens, 2004–2006*. US Geological Survey Professional Paper 1750. Reston, VA: U.S. Geological Survey.

Short, C. Brant. 1989. *Ronald Reagan and the Public Lands: America's Conservation Debate, 1979–1984*. College Station: Texas A&M University Press.

Siebert, Lee. 2005. "Blown Away." *Natural History* 114 (8): 50–55.

Sieh, Kerry, and Simon LeVay. 2005. *Mount St. Helens: Photographs by Frank Gohlke*. New York: The Museum of Modern Art.

Snyder, Gary. 2004. *Danger on Peaks*. Washington, DC: Shoemaker & Hoard.

Sorenson, John H. 1982. In *Status of Volcanic Prediction and Emergency Response Capabilities in Volcanic Hazard Zones of California*, edited by Roger C. Martin and James F. Davis, 153–67. Sacramento: California Department of Conservation.

Spies, Thomas A., and Sally L. Duncan, eds. 2009. *Old Growth in a New World: A Pacific Northwest Icon Reexamined*. Washington, DC: Island Press.

Stegner, Wallace. 1999. *Marking the Sparrow's Fall: The Making of the American West*. New York: Holt.

Stone, Brad. 2013. *The Everything Store: Jeff Bezos and the Age of Amazon*. New York: Little Brown.

Taylor, Quentin. 2005. "Money and Politics in the Land of Oz." *The Independent Review* (Winter).

Thomas, Jack Ward, Jerry F. Frankin, John Gordon, and K. Norman Johnson. 2006. "The Northwest Forest Plan: Origins, Components, Implementation Experience, and Suggestions for Change." *Conservation Biology* 20 (2): 277–87.

Thompson, Dick. 1980. *Volcano Cowboys: The Rocky Evolution of a Dangerous Science*. New York: St. Martin's Press.

Thoreau, Henry David. 1984. *The Maine Woods*. Boston: Ticknor & Fields.

Tilling, Robert I. 1989. "Volcanic Hazards and Their Mitigation: Progress and Problems." *Reviews of Geophysics* 27 (2): 237–69.

——. 2000. "Mount St. Helens 20 Years Later: What We've Learned." *Geotimes* 45 (May): 14–19.

Tilling, Robert I., Lyn Topinka, and Donald A. Swanson. 1990. *Eruptions of Mount St. Helens: Past, Present, and Future*. Revised edition. Washington, DC: US Department of the Interior.

Twining, Charles E. 1985. *Phil Weyerhaeuser: Lumberman*. Seattle: University of Washington Press.

United States Congress. 1982. *Mount St. Helens national volcanic area: joint hearings before the Subcommittee on Forests, Family Farms, and Energy of the Committee on Agriculture and the Subcommittee on Public Lands and National Parks of the Committee on Interior and Insular Affairs, House of Representatives, Ninety-seventh Congress, second session, on H.R. 5281, H.R. 5773, H.R. 5787, March 11, 1982, Washington, D.C., April 3, 1982, Vancouver, Wash*. Washington, DC: Government Printing Office.

Urie, Steve. 2012. *Rock Island: An American History*. Reno, NV: Small Pond Publishing Co.

Vallance, J. W., C. A. Gardner, W. E. Scott, R. M. Iverson, and T. C. Pierson. 2010. "Mount St. Helens: A 30-Year Legacy of Volcanism." *Eos* 91 (19): 169–72.

Van Syckle, Edwin. 1980. *They Tried to Cut It All: Grays Harbor, Turbulent Years of Greed and Greatness*. Seattle: Pacific Search.

Verhovek, Sam Howe. 2010. *Jet Age: The Comet, the 707, and the Race to Shrink the World*. New York: Avery.

Voight, Barry. 1981. "Time Scale for the First Moments of the May 18 Eruption." In *The 1980 Eruptions of Mount St. Helens, Washington*, edited by Peter W. Lipman and Donal R. Mullineaux, 69–86. Geological Survey Professional Paper 1250. Washington, DC: Government Printing Office.

——. 2000. "Structural Stability of Andesite Volcanoes and Lava Domes." *Philosophical Transactions of the Royal Society of London A* 358 (1770): 1663–703.

Voight, Barry, Harry Glicken, R. J. Janda, and P. M. Douglass. 1981 "Catastrophic Rockslide Avalanche of May 18." In *The 1980 Eruptions of Mount St. Helens, Washington*, edited by Peter W. Lipman and Donal R. Mullineaux, 347–77. Geological Survey Professional Paper 1250. Washington, DC: Government Printing Office.

Waitt, Richard. 2014. *In the Path of Destruction: Eyewitness Chronicles of Mount St. Helens*. Pullman: Washington State University Press.

Warren, Debra D. 2011. *Harvest, Employment, Exports, and Prices in Pacific Northwest Forests, 1965–2010*. Portland, OR: United States Department of Agriculture, Forest Service, Pacific Northwest Research Station.

Weyerhaeuser, F. K. 1951. *Trees and Men*. New York: Newcomen Society of North America.

Weyerhaeuser, Louise. 1940. *Frederick Weyerhaeuser: Pioneer Lumberman*. Minneapolis: McGill Lithograph Company. [Weyerhaeuser family private publication]

White, Richard. 1980. *Land Use, Environment, and Social Change: The Shaping of Island County, Washington*. Seattle: University of Washington Press.

——. 2011. *Railroaded: The Transcontinentals and the Making of Modern America*. New York: Norton.

Williams, Gerald W. 2009. *The U.S. Forest Service in the Pacific Northwest: A History*. Corvallis: Oregon State University Press.

Williams, Gerald W., and Char Miller. 2005. "At the Creation: The National Forest Commission of 1896–97." *Forest History Today* (Spring/Fall): 32–40.

Williams, Michael. 1989. *Americans and Their Forests: A Historical Geography*. Cambridge: Cambridge University Press.

——. 2006. *Deforesting the Earth: From Prehistory to Global Crisis*. Chicago: University of Chicago Press.

Winchester, Simon. 2003. *Krakatoa: The Day the World Exploded: August 27, 1883*. New York: HarperCollins.

Wood, Charles R. 1968. *The Northern Pacific: Main Street of the Northwest*. Seattle: Superior Publishing Company.

Wood, Nathan J., and Christopher E. Soulard. 2009. *Community Exposure to Lahar Hazards from Mount Rainier, Washington*. US Geological Survey Scientific Investigations Report 2009–521. Reston, VA: US Geological Survey.

ILLUSTRATION CREDITS

Maps in text are made by Matt Stevenson, CORE GIS.

Part 1: *Mount St. Helens before the 1980 eruption, with logged Weyerhaeuser land and Fawn Lake in the foreground.* U.S. Geological Survey Department of the Interior / USGS U.S. Geological Survey.

Part 2: *The largest tree Weyerhaeuser ever cut, just to the northwest of Mount St. Helens.* Longview Room Collection. Longview Public Library.

Part 3: *Mount St. Helens rising above Spirit Lake.* "Mount St. Helens, Spirit Lake, boaters," 1940–1980, Josef Scaylea, State Library Photograph Collection, 1851–1990, Washington State Archives, Digital Archives, http://www.digitalarchives.wa.gov.

Part 4: *Mount St. Helens at 8:33 a.m. on Sunday, May 18, 1980.* © 1980 Keith Ronnholm.

Part 5: *Station wagon hit by a fallen tree at the edge of the blast zone.* U.S. Geological Survey Department of the Interior / USGS U.S. Geological Survey.

Part 6: *Susan Saul hiking above Spirit Lake in the Mount St. Helens National Volcanic Monument.* Photograph by the author.

Part 7: *The Pumice Plain to the north of Mount St. Helens in 2005.* U.S. Geological Survey Department of the Interior / USGS U.S. Geological Survey.

INSERT

Page 1, top: *Painting by Canadian artist Paul Kane of the 1847 eruption of Mount St. Helens.* With permission of the Royal Ontario Museum © ROM.

Page 1, bottom: *Mount St. Helens and Spirit Lake before the 1980 eruption.* U.S. Geological Survey Department of the Interior / USGS U.S. Geological Survey.

Page 2, top left: *Frederick Weyerhaeuser, who built a lumber empire from a small sawmill on the banks of the Mississippi River.* Rock Island Company.

Page 2, top right: *James J. Hill, his next-door neighbor in St. Paul.* Minnesota Historical Society.

Page 2, bottom left: *Gifford Pinchot, at about the time he served as secretary for the National Forest Commission.* US Forest Service at Grey Towers National Historic Site.

Page 2, bottom right: *President Theodore Roosevelt.* 1902, Library of Congress.

Page 3, top: *The newly formed crater on top of the volcano in late March 1980.* U.S. Geological Survey Department of the Interior / USGS U.S. Geological Survey / Photo by D. Frank.

Page 3, middle: *Geologist Dave Johnston collecting samples from the crater in April 1980.* U.S. Geological Survey Department of the Interior / USGS U.S. Geological Survey / Photo by R.P. Hoblitt.

Page 3, bottom: *The Coldwater I station on the edge of a ridge eight miles from the volcano.* U.S. Geological Survey Department of the Interior / USGS U.S. Geological Survey.

Page 4, top left: *Dave Johnston at Coldwater II the day before the May 18 eruption.* U.S. Geological Survey Department of the Interior / USGS U.S. Geological Survey.

Page 4, top right: *An ash-covered Mount St. Helens from Coldwater II on May 17, the day before the eruption.* U.S. Geological Survey Department of the Interior / USGS U.S. Geological Survey / Photo by Henry Glicken.

Page 4, bottom: *Mount St. Helens erupting on Sunday, May 18, 1980.* U.S. Geological Survey Department of the Interior / USGS U.S. Geological Survey / Photo by Robert Krimmel.

Page 5: *The blast cloud on the east side of the volcano.* © 1980 Keith Ronnholm.

Page 6, top: *Fawn Lake after the eruption, where John and Christy Killian were camping.* U.S. Geological Survey Department of the Interior / USGS U.S. Geological Survey / Photo by Lyn Topinka.

Page 6, bottom: *A Weyerhaeuser company crummy submerged in a mudflow from the volcano.* U.S. Geological Survey Department of the Interior / USGS U.S. Geological Survey.

Page 7, top: *Rescuers looking at the remains of photographer Reid Blackburn in his ash-filled car at Coldwater I.* U.S. Geological Survey Department of the Interior / USGS U.S. Geological Survey.

Page 7, middle: *Ralph Killian looking for his son at Fawn Lake the year after the eruption.* Chris Johns / Seattle Times.

Page 7, bottom: *The Green River valley looking west toward the area where Clyde Croft and Al Handy were camping on the morning of the eruption.* Courtesy of Greg Wahlsteg.

Page 8, top: *Satellite photograph of Mount St. Helens taken in the spring of 2015.* Courtesy of NASA.

Page 8, bottom: *Mount St. Helens in 1982, with a new volcanic cone rising from its crater.* U.S. Geological Survey Department of the Interior / USGS U.S. Geological Survey / Photo by Lyn Topinka.

INDEX

Note: Italic page numbers refer to charts, maps, and illustrations. "Insert" before page numbers refers to insert pages.

ABOUT THE AUTHOR

PHOTOGRAPH BY DANI WEISS

STEVE OLSON is the author of *Mapping Human History: Genes, Race, and Our Common Origins*, which was nominated for the National Book Award and received the Science-in-Society Award from the National Association of Science Writers, and *Count Down: Six Kids Vie for Glory at the World's Toughest Math Competition*, which was named a best science book of the year by *Discover* magazine. His most recent book, co-written with Greg Graffin, is *Anarchy Evolution: Faith, Science, and Bad Religion in a World Without God*. Except for three years as a writer in the White House science office, he has been a freelance writer and editor since 1979. He has written for the *Atlantic Monthly, Science, Smithsonian, Seed*, the *Washington Post*, the *Los Angeles Times, Scientific American, Wired*, the *Yale Alumni Magazine*, the *Washingtonian, Slate, Astronomy, Science 82–86*, and many other magazines. He now lives in Seattle, Washington.